D1385018

Invisible Worlds

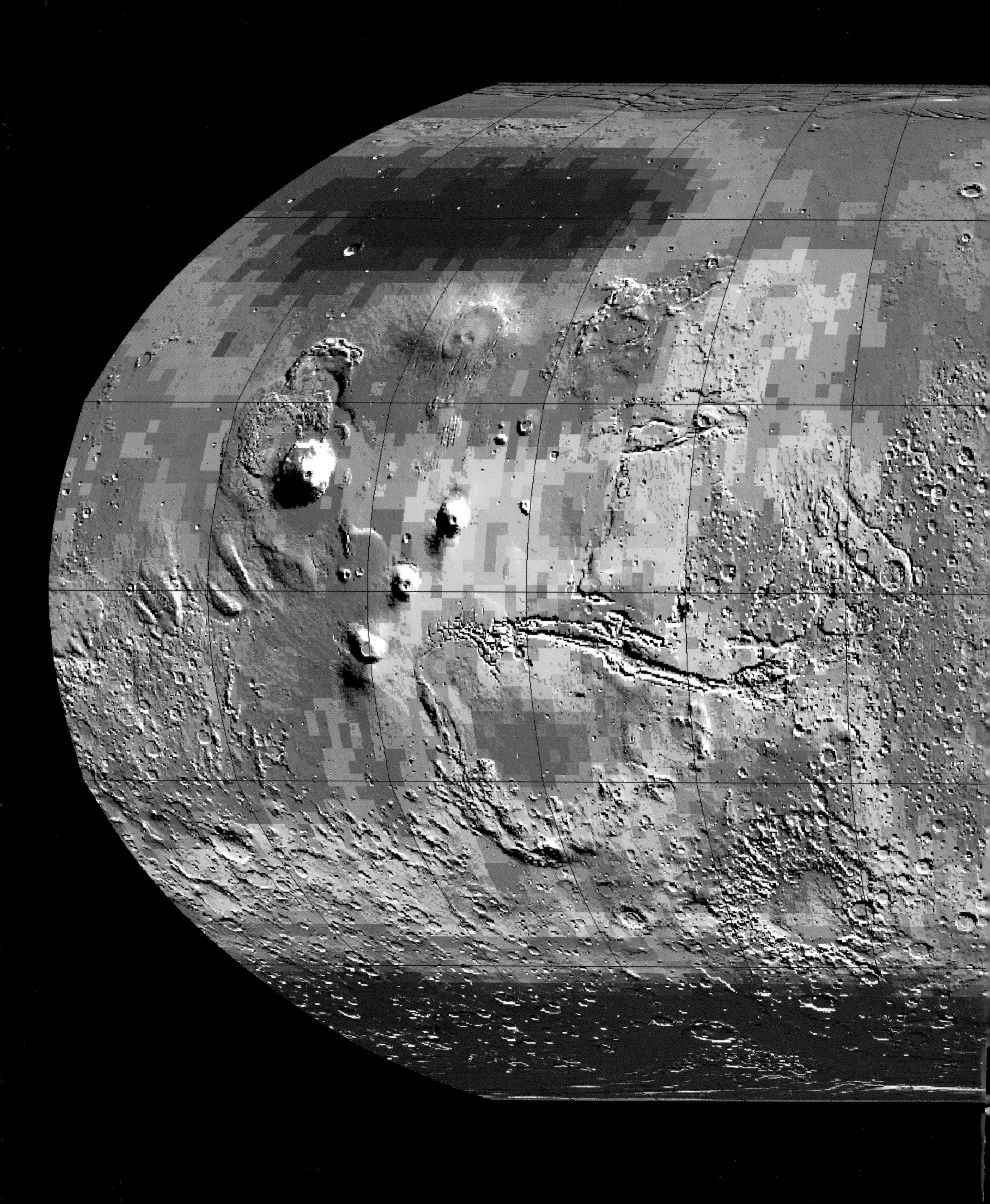

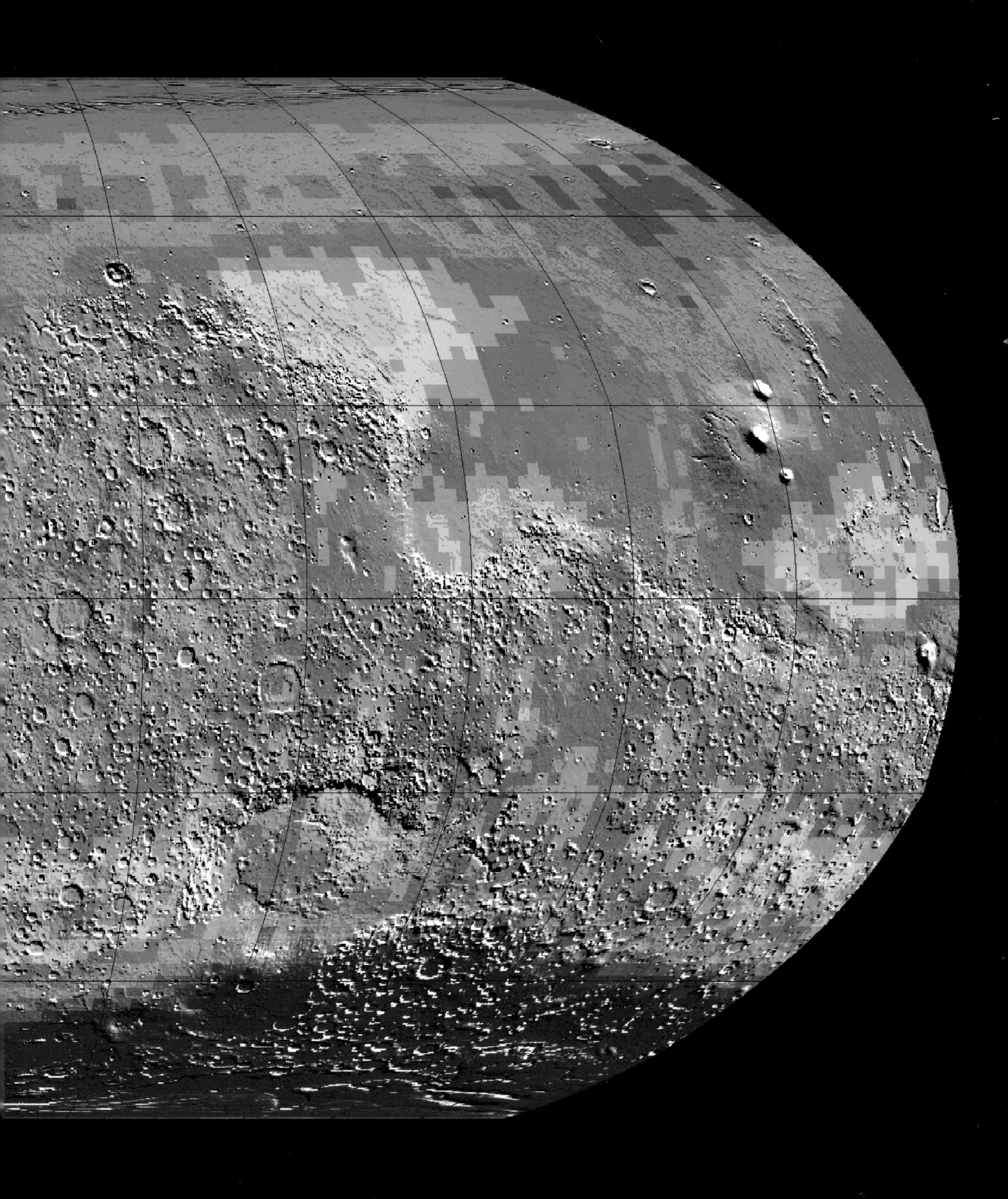

Pages 2–3
A neutron spectrometer aboard NASA's Mars Global Surveyor spacecraft mapped the distribution of hydrogen atoms in the planet's topsoil. The hydrogen (in blue) is bound up in huge reserves of water ice.

Pages 4–5
The blood system in a tissue sample is filled with a plastic resin. Surrounding skin and muscle cells are chemically dissolved away, and an electron microscope can then capture the fragile tracery of vessels.

Pages 6–7
A space shuttle camera records cities and towns at night. Scans taken during several orbits were combined into one seamless image, and light levels were calibrated to give an accurate impression of relative brightness.

Invisible Worlds

Exploring the Unseen

WEIDENFELD & NICOLSON

Contents

Preface
invisible worlds

A few years ago I went with friends to a well-known tourist destination on the south coast of England known as Beachy Head. It is not the beaches so much as the tall, sheer, white chalk cliffs, capped by rich green pastures that give the site its unique character. On this particular day the Sun was blazing down from a beautiful sky of rich deep blue, decorated with occasional cotton-wool clouds. Wild flowers advertised themselves as vivid specks of red and yellow among the greener-than-green grass, and seagulls wheeled and screeched overhead. It was a perfect summer's afternoon. While my friends expressed understandable delight at the loveliness of it all, I suddenly experienced a feeling that I can only compare, as a physical sensation, to vertigo: a flash of near-terror at some infinitely deep chasm into which I might fall. I had to sit down for a moment and catch my breath.

But it was not the height of the cliffs that had appalled me. No – it was the onrush of an immense sense of utter blindness to the wonderful scenery. The blue of the sky, the green of the grass – even the fiery disc of the Sun – all seemed like the most trivial curtains of illusion drawn across the truth of the world. My eyes were working perfectly well, but I felt as if I could not actually *see* anything at all. It was as though the cliff, the sea, the huge expanse of sky were taunting me with their inaccessibility. The colours were like a smoke-screen hiding what I knew had to be a much vaster and more subtle 'true' version of this magnificent scene. Birds wheeling over my head sensed magnetic fluxes; insects responded to polarised light and ultraviolet reflections in the grass and flowers; and all around us blasts of radio, infrared and ultra-violet energies from the Sun were blazing so intensely that our senses should have been overwhelmed. Instead, we were blind to almost all of it.

It was a fleeting sensation and no more, but it did spark in me a desire to try to think about what the world 'really' looks like, and what it 'really' is. What we see, even on the brightest summer's afternoon, is just the smallest fraction of what is actually out there. Beyond the familiar rainbow of visible colours, nature conceals a far wider spectrum of invisible energies. There is a hidden universe whose depth and beauty were entirely unsuspected until recent times. From the subatomic to the cosmic, we have discovered patterns and forces, shapes and objects that were quite literally beyond the vision of our naturally evolved eyes until we learned how to make pictures outside the restricting limits of daylight. This book explores how those pictures are created, and how we can interpret them.

Along the way, we delve into the hearts of atoms and into the tiniest molecules of life, and witness monstrous predators lurking in a speck of dust. We examine the delicate workings of our minds and bodies with scanners that defy the most outlandish predictions of science fiction. We see the dramatic hidden life of the Earth, as revealed by sound waves, radio waves, infrared and ultraviolet. Then we visualise the massive forces and energies that shape the planets, the Sun, and other stars and galaxies, all of which are normally quite invisible to us. And with the power of computers we dive into the abstractions of modern mathematics and see hauntingly beautiful 'objects' that do not exist, yet which tell us – perhaps – some profound truths about how nature works. We can glimpse the ultimate textures of reality at all scales, from the smallest constituents of matter to the possible shape of the entire known universe. We also discover an intriguing borderland where science and art appear to converge.

Some fascinating questions arise. Is it possible to make a picture of an atom? Who adds the artificial colours to an electron microscope scan of a flea, and why? What, if anything, do positron-emission brain scans of the mind actually tell us about consciousness? When is a gamma ray map of Mars showing buried ice-water not all that it seems? Can we really 'see' proof, written in the subtlest microwave energies, of the Big Bang that created our universe? Are we daily encountering hitherto unknown patterns in nature, or do the latest scanning technologies – our new electronic eyes – reveal only what we already expect to find?

If this book succeeds in its purpose, we can begin to wonder what a bright summer's afternoon might really look like – if only we had the eyes to see.

Introduction
the hidden rainbow

A VAST SPECTRUM of invisible light energies exists beyond the reach of unaided human senses. They dominate the science and technology of the modern world, and we have discovered how to translate them into light that we can see.

Waves in the electromagnetic spectrum vary in size, from radio waves the length of buildings, to gamma rays smaller than the nucleus of an atom. The waves take the form of intertwined electric and magnetic fields, travelling together through space while oscillating perpendicularly to each other. The shorter the wavelength, the higher the energy. At the gamma extreme of the spectrum, the wavelengths are so short and punchy they can smash into solid matter and wreak havoc. At the other end, the longest and most languorously-spaced radio waves have very little energy and barely interact with the solid world.

The fundamental carrier of electromagnetic energy is the photon, a particle of light that has no mass but can convey energy: a concept that is very tricky to reconcile with the notion of light as spread-out waves propagating through space. All electromagnetic radiation is 'light', even though we can only see a limited range of it through our eyes. Photons are emitted by atoms when electrons – particles that orbit the atomic nucleus – are kicked into a higher, more energetic orbit as a result of being heated or energetically boosted in some way, before falling back down to their original lower orbit and shedding their excess energy in the form of photons. The more energy that is fed into the electrons to begin with, the higher an orbit they climb to, and the higher the energy of the photons emitted when the electrons fall back. However, energies can only be absorbed or emitted in discrete, indivisible packets called 'quanta'. That is essentially why different atomic elements, with distinct arrangements of electrons around their nuclei, give off characteristic colours of light at very specific wavelengths (photon energies) when they are burned.

Low-energy photons, such as radio, behave like waves, while higher-energy photons, such as X-rays, behave like particles. Here is where we encounter a very puzzling feature of light. Is a beam of light in fact a continuous wave, or is it a collection of discrete particles? Wavelengths are infinitely variable, while photon energies are not. Modern physics has confirmed that both pictures are valid depending on the way we use, or measure, light. Scientists and instrument designers work with whichever description of light they need at the time: undulating waves or bullet-like photons. All the ways of thinking about electromagnetic radiation can be related to the others in precise mathematics. Wavelength always equals the speed of light divided by the frequency; and the energy contained in photons is always proportional to the frequency of a wave, with the proportion governed by Planck's constant (named in honour of the physicist who discovered it, heralding the birth of quantum theory). The Planck number, and the speed of light, appear to be immutable fixed values set into the laws of nature.

Wavelength = the speed of light divided by the frequency.
Energy = Planck's constant x the frequency.

We have to keep the particle and wave analogies in mind simultaneously, like 'doublethink' in Orwell's *1984*. It is hard to imagine a wave compact enough to collide with a single atomic electron, leaving all the other neighbouring electrons unaffected, but that is what X-ray waves can do. It is even stranger to think of subatomic photons smearing themselves through space to become waves the size of buildings, but that is what radio photons can do. None of it makes intuitive sense, but the mathematical descriptions of these phenomena are among the most reliable tools of the human intellect, and the technologies based on them dominate our everyday lives, from FM radio and mobile phones, to microwaves, sunlamps, hospital X-ray machines and astronomical telescopes. It is just that asking what electromagnetic radiation actually *is* might drive us insane.

Introduction
the hidden rainbow

The sentence 'Grant eXpects Unanimous Votes In Movie Reviews' may help in remembering the order of the electromagnetic spectrum. It is a mnemonic for G, X, U, V, I, M, R: Gamma rays, X-rays, Ultraviolet, Visible, Infrared, Microwave and Radio. (There do remain some areas of overlap.)

Wavelengths are usually expressed in metres, or fractions of a metre, written as powers of ten for convenience. The wavelength of visible light is usually expressed in nanometres (billionths of a metre). Frequencies are

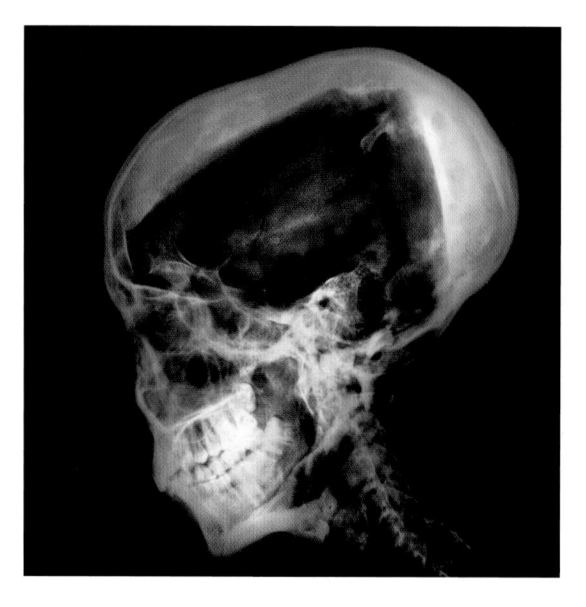

X-rays have been harnessed in medicine for 100 years. Short wavelength radiation passes through the body, and bones and denser tissues leave a shadow on a photographic plate exposed to the rays.

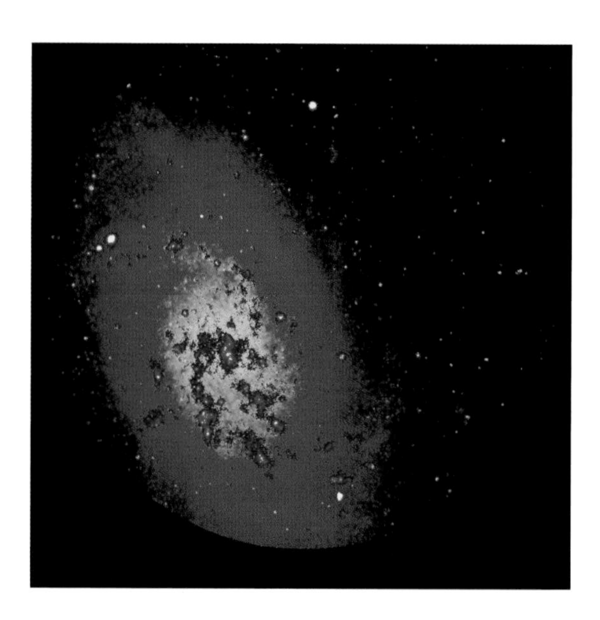

Ultraviolet instruments aboard space telescopes are revealing the deeper structures and life cycles of distant galaxies. Visible light from stars is only a small fraction of the electromagnetic energies they emit.

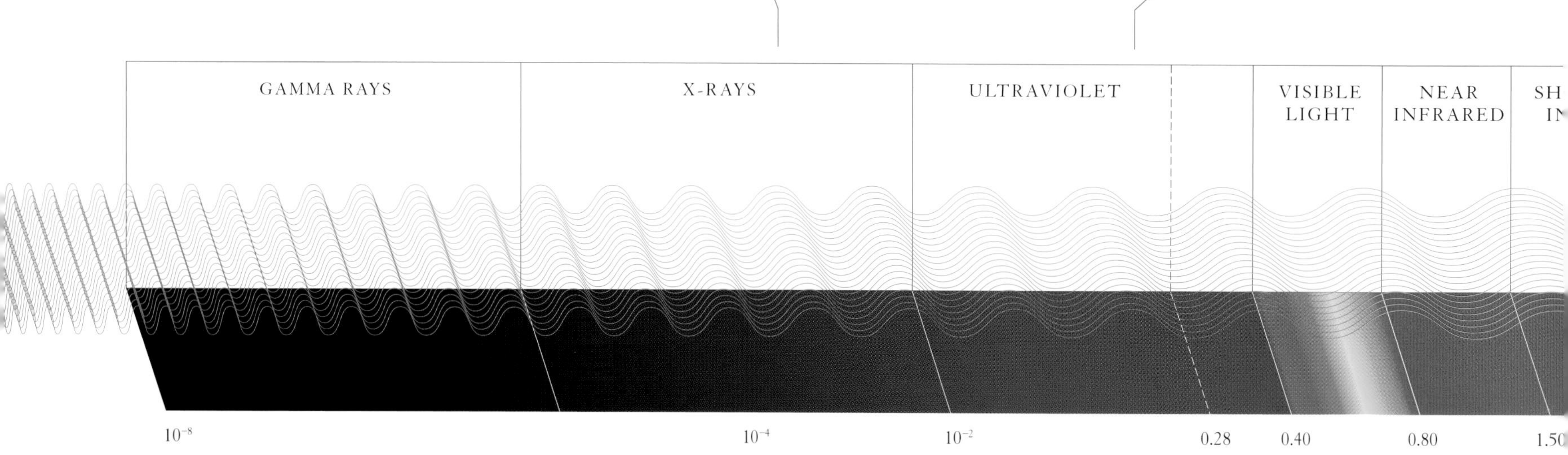

described as the number of wave oscillations per second (hertz). X-rays and gamma rays are usually expressed in terms of their energy, which is proportional to their frequency.

This book is inspired by techniques which make images in non-optical wavelengths. But some other 'invisible' phenomena, such as gravity and sound waves, are also included – along with certain patterns in nature that cannot easily be pictured with instruments.

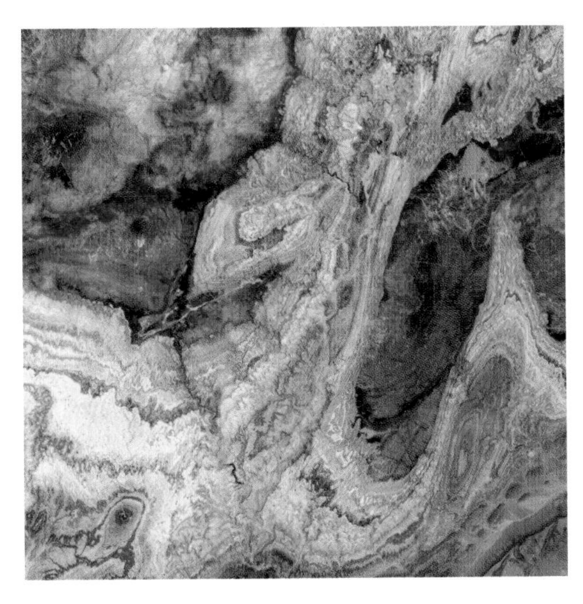

Orbiting astronauts see the Earth as a beautiful blue ball illuminated by sunlight. Robotic instruments working at infrared and microwave frequencies are able to provide an entirely different view of the planet.

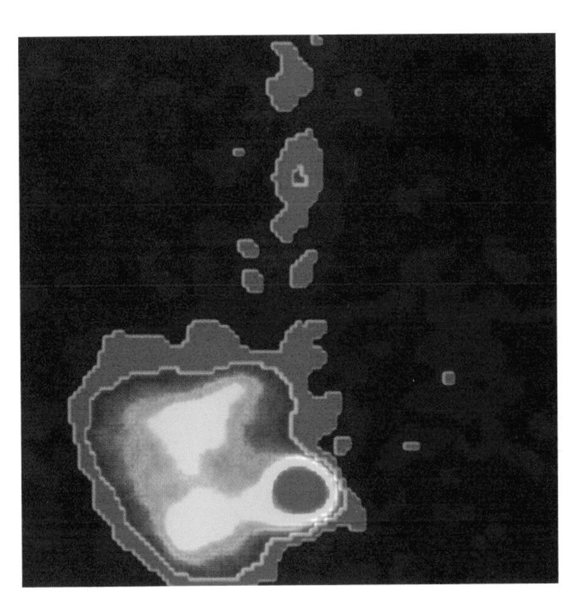

We are accustomed to 'hearing' radio broadcasts, but astronomers have learned how to 'see' natural radio waves that emanate from deep space. The cosmos is teeming with energetic radio sources.

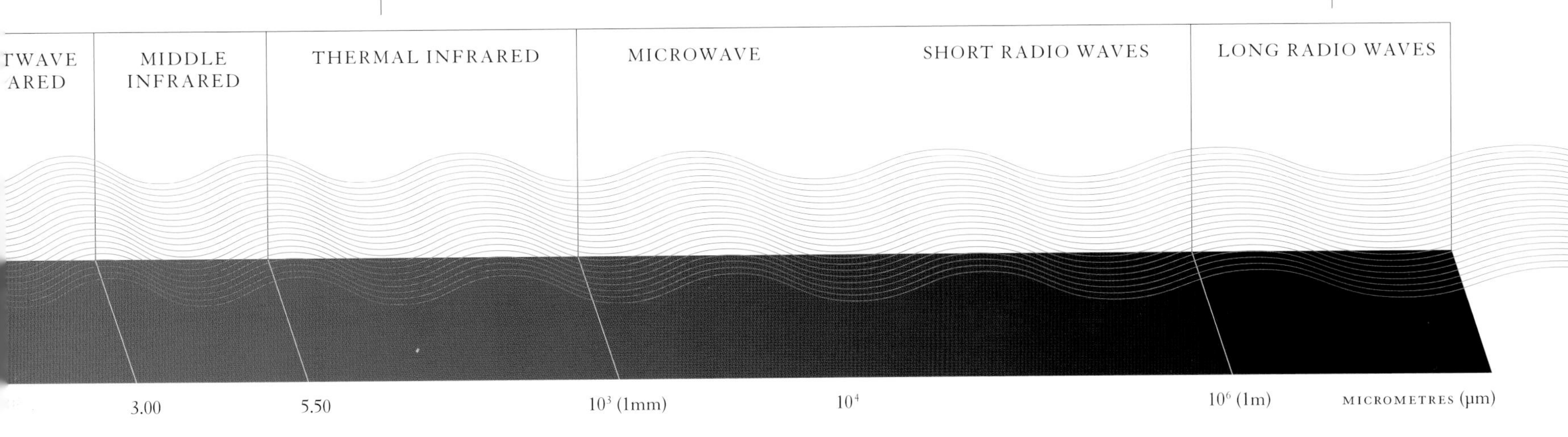

1

A cloud of uncertainty

By harnessing powerful energies, our largest instruments produce images of the smallest things: atoms, electrons, protons, neutrons and even smaller particles. At the molecular scale, biologists also use high energies to learn about the chemistry of life. From the smashing of atoms to the diffraction of X-rays skimming through a crystal of DNA, the deepest patterns of nature's building blocks are revealed to thrilling effect. But these beautiful, eerily abstract images also prove that the 'stuff' our world is made from is very strange indeed.

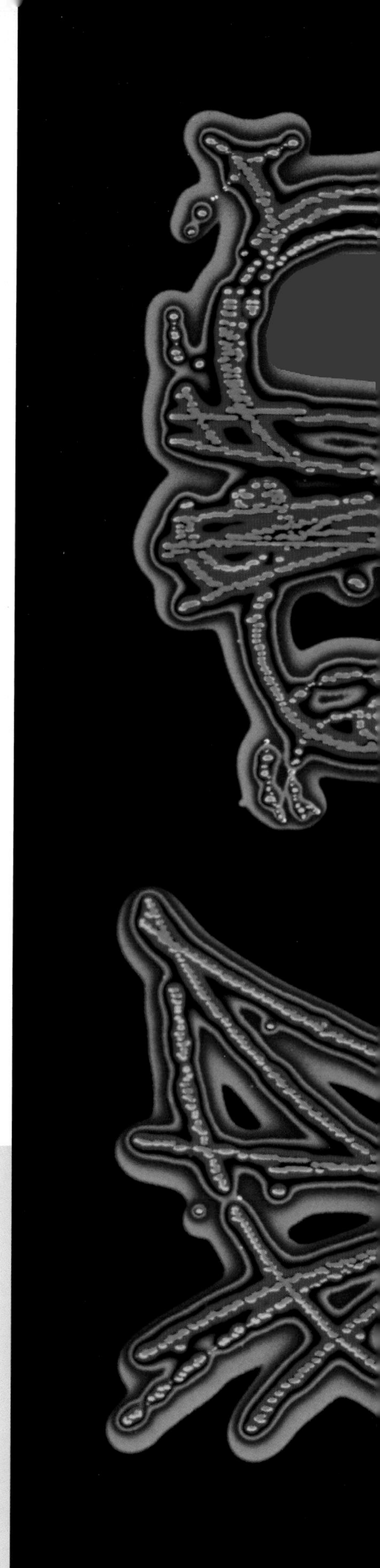

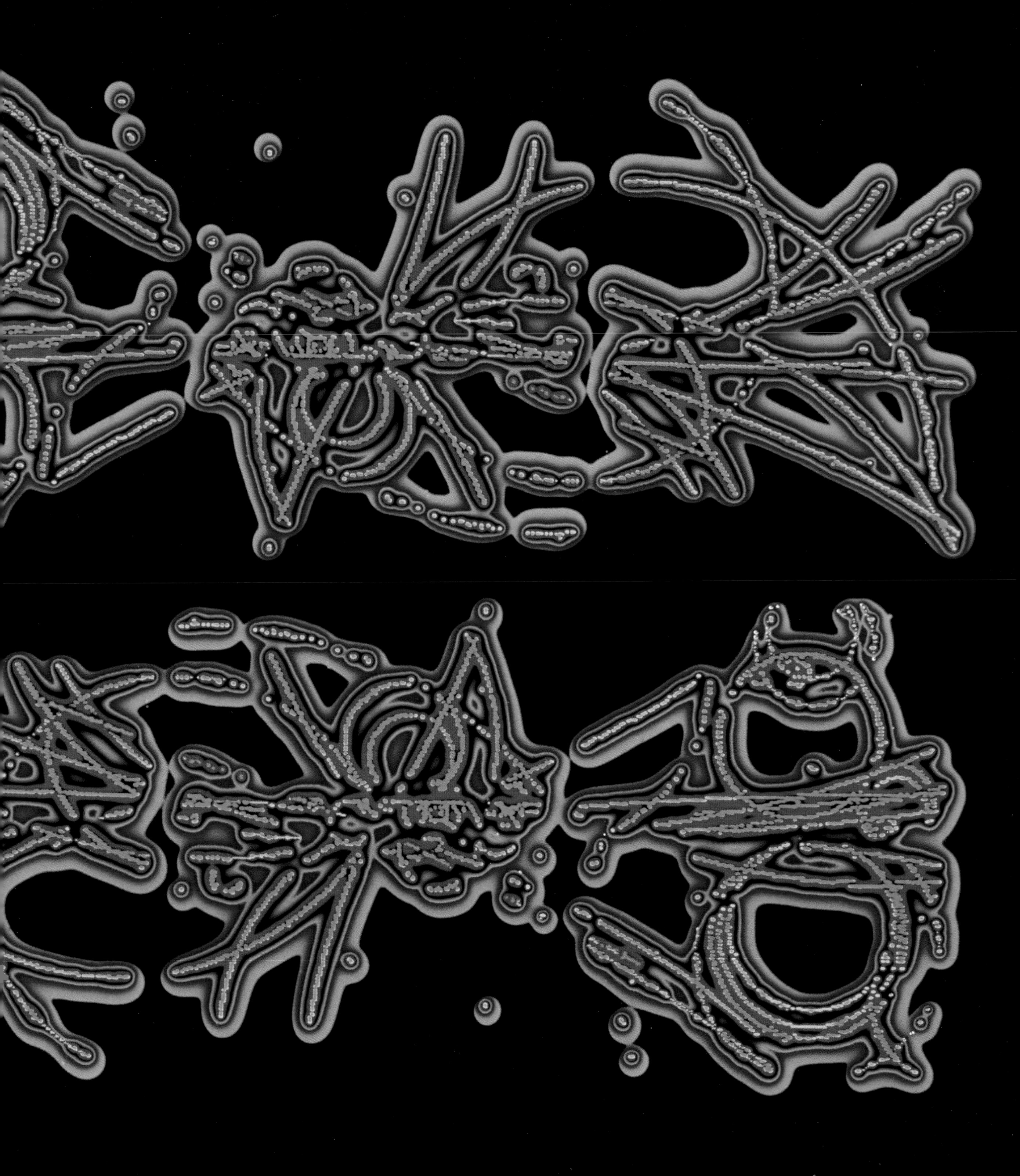

First images of the invisible

iron filings reveal magnetic fields

Isaac Newton described a universe where gravity was an invisible force that caused masses (apples, planets, stars, galaxies) to pull towards each other, even when separated by a void. He was not able to explain what gravity actually was. Nowadays, that mystery remains more or less intact, although we have devised an alternative way of describing gravity. It is a curvature of space, a 'field' caused by the distorting presence of mass and energy in space-time.

Today we are able to describe many things, from magnetism to electricity, gravity and subatomic phenomena, in terms of fields. Magnetism, like gravity, is a field; except it can repel other magnetic objects as well as drawing them in. Everyday electrical devices, motors, televisions, computers, electrical wiring in homes and the power lines that supply them all create magnetic fields. Even the human body produces fields, generated by chemical reactions within cells and the flow of ions in the nervous system. The Earth has powerful magnetic and gravitational fields, as do most planets and moons, and all stars and galaxies. The energies of the electromagnetic spectrum are also amenable to 'field' descriptions, as well as 'wave and particle' theories.

Are fields real? In 1850, Michael Faraday made 'iron filing diagrams' to illustrate that the empty space immediately surrounding a magnet had structure. He scattered iron filings on to sheets of paper impregnated with wax. Tapping the paper caused the filings to arrange themselves neatly along lines of force, surrounding the magnet like a ghostly aura. A gentle heating of the wax then glued the filings into place, and the first pictures of invisible fields had been created. The results, preserved to this day at the Royal Institution in London, seemed to prove the existence of magnetic fields; but the picture is literally not so straightforward. What we actually see are the products of the iron filings' interactions with a nearby magnet. There is no proof that magnetic fields (or gravitational fields for that matter) exist as entities in their own right, independently of the things that are affected by them. Faraday's 150-year-old wax sheets still present us with a profound mystery. Not only can we not be sure of what we are seeing, we cannot even be sure if it is really there in the first place. Science tends to describe one thing in terms of its *effect* on another. Getting to the nub of any one thing in itself is often profoundly difficult.

PRECEDING PAGES
Subatomic particles flying across a chamber at near-light speeds leave trails of gas bubbles in a tiny tank of liquid hydrogen. The microscopic traces are enhanced electronically in this image from the CERN laboratory near Geneva.

FACING PAGE
A modern recreation of Faraday's experiments shows iron filings clustered around a pair of bar magnets, which each have a north and a south pole of field orientation. Opposite poles attract each other. In this case, two like poles create a zone of repulsion between the magnets.

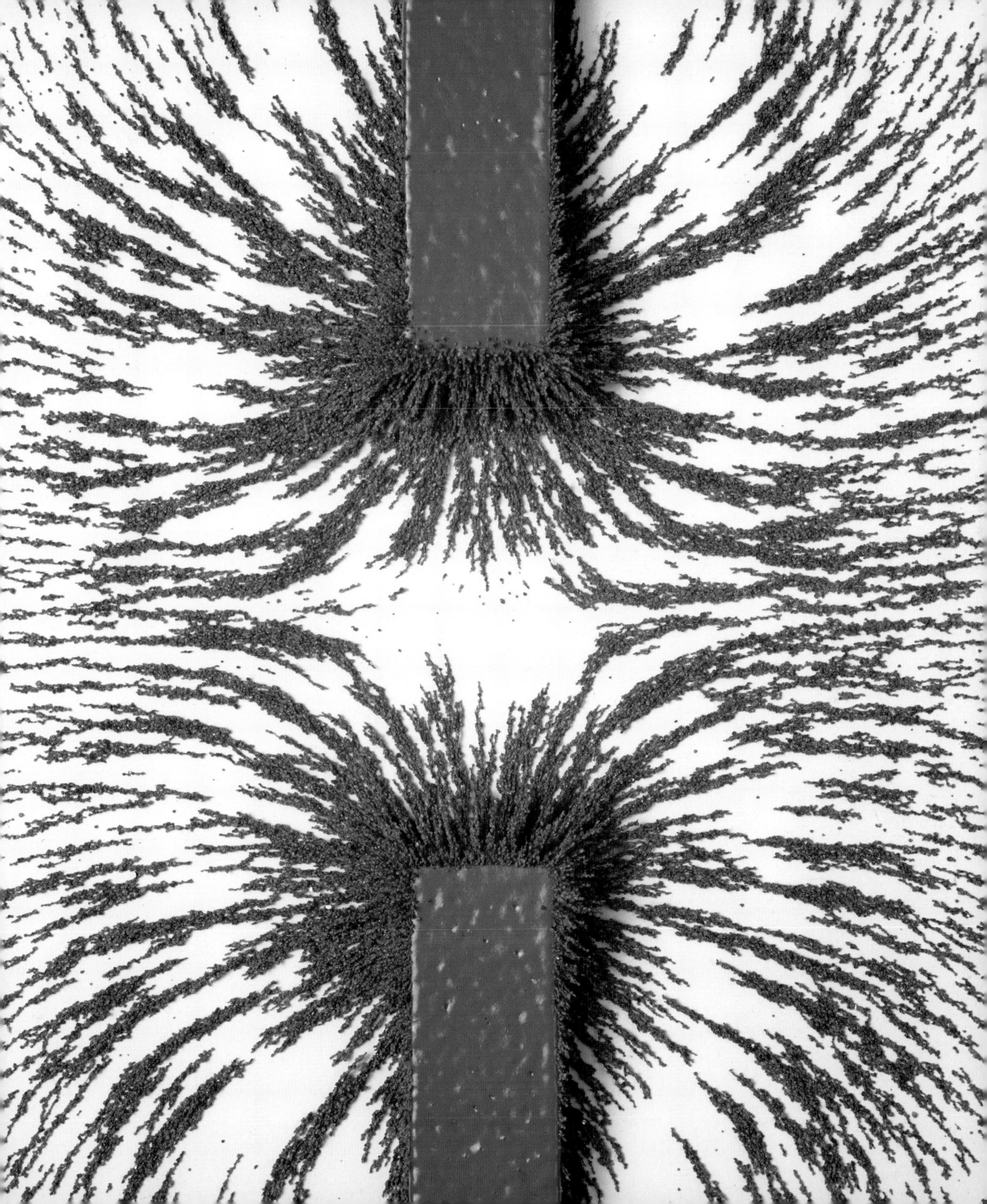

More than meets the eye

early encounters with unknown radiations

In 1873, the Scottish physicist James Clerk Maxwell expanded on Michael Faraday's earlier work with electricity and magnetic lines of force. He showed that a few relatively simple mathematical equations could express the behaviour of electric and magnetic fields; and he proved that an oscillating electric charge produces a combined electromagnetic field which propagates through space at the speed of light. Maxwell realised that light itself had to be an electromagnetic phenomenon, and he also concluded that, since electromagnetic energies could oscillate at any frequency, then visible light must constitute only a tiny portion of a vastly wider electromagnetic spectrum. It was a brilliant deduction that would one day transform all of science.

Heinrich Hertz, a German physicist who was born in 1857 (and died of blood poisoning at the tragically early age of 37), applied Maxwell's theories to the production and reception of 'electric waves'. In recognition of his work, the unit of frequency is even today named the hertz. In 1888, in a corner of his physics classroom at the Karlsruhe Polytechnic in Berlin, Hertz generated radio waves with an electrical sparking machine, and then detected them with a similar type of circuit on the other side of the room. He had discovered how to make the electric and magnetic fields detach themselves from the wires of his apparatus and travel freely through space. Hertz's students were impressed, but he thought his discoveries were no more practical than Maxwell's. 'It's of no use whatsoever,' he said. 'We just have these mysterious electromagnetic waves that we cannot see with the naked eye. But they are there.'

A precociously intelligent young teenager happened to read about Hertz's work later that year, while on vacation in the Alps. He rushed back home to Italy to see if he could use Hertz's discoveries as part of a long-distance signalling device. His name was Guglielmo Marconi. By 1895 he had invented the first practical system for radio telegraphy, capable of sending and receiving pulsed signals at a range of more than a kilometre. By the turn of the century he was transmitting between different continents. Invisible electromagnetic energies were no longer just theoretical curiosities; they were about to change the world.

Two overlapping sine waves on an oscilloscope monitor give some impression of the form of radio waves. The frequency (number of waves per second) usually stays fixed so that receivers can be tuned to catch a particular signal.

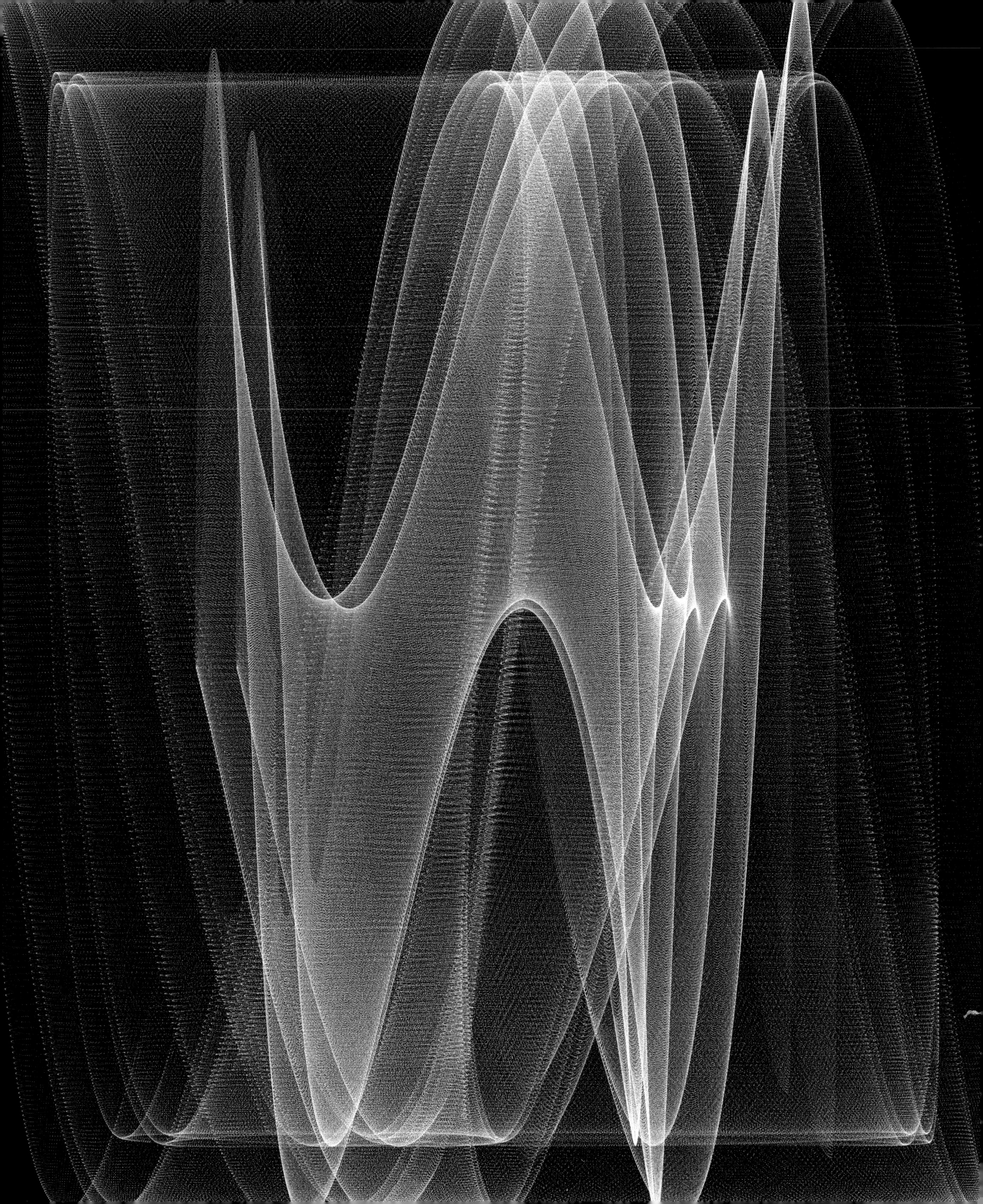

Lively molecules

the lessons of Brownian motion

In 1827, the Scottish botanist Robert Brown was examining pollen grains under a simple microscope. He had suspended fine grains in water, and he noted how they seemed to move about, tracing random zigzag paths across his microscope's field of view. At first, he concluded that the movement of each grain 'arose neither from currents in the fluid, nor from its gradual evaporation, but belonged to the particle itself.' Other observers enthusiastically concluded that Brown had witnessed a fundamental 'life force' animating the tiniest pieces of biological matter. His initial findings caused great excitement.

Brown was a conscientious scientist. Even as he prepared to publish his results, he revised his text to warn that he had seen a similar motion among pollen grains that he had preserved in alcohol many months before, which surely must have been lifeless by the time he put them under the microscope. Of course, there was a slim chance that pollen was harder to kill than he had assumed, so one more experiment was needed to remove that ambiguity. He ground inorganic mineral samples into powders and suspended them in water. Again, he saw random movements through his microscope. If there was some kind of a life force at work here, it almost certainly was not being exerted by the grains themselves: it had to be something in the water.

This is an experiment that any of us can do with a simple child's microscope. It demonstrates the existence of water molecules – or more specifically, it proves they are in constant motion. We cannot see them directly through an ordinary microscope because they are far too small, but we can see how their constant, random buffeting of grains suspended in the water pushes the grains about.

Actually, no. What we 'see' is grains scuttling about against a clear background, like little bugs. Brown's experiment tells us that the appearances of things are not necessarily the same as their meanings. Today we have created scientific instruments that are to Brown's little brass microscope what a stadium floodlight is to a candle. We still have to be wary of what we think we see through them.

Pollen grains from a larch tree, suspended in water and photographed through a microscope, give some impression of what Brown would have studied. Even today, with our knowledge of Brownian motion, it is intriguing to watch the grains moving around as if at their own volition.

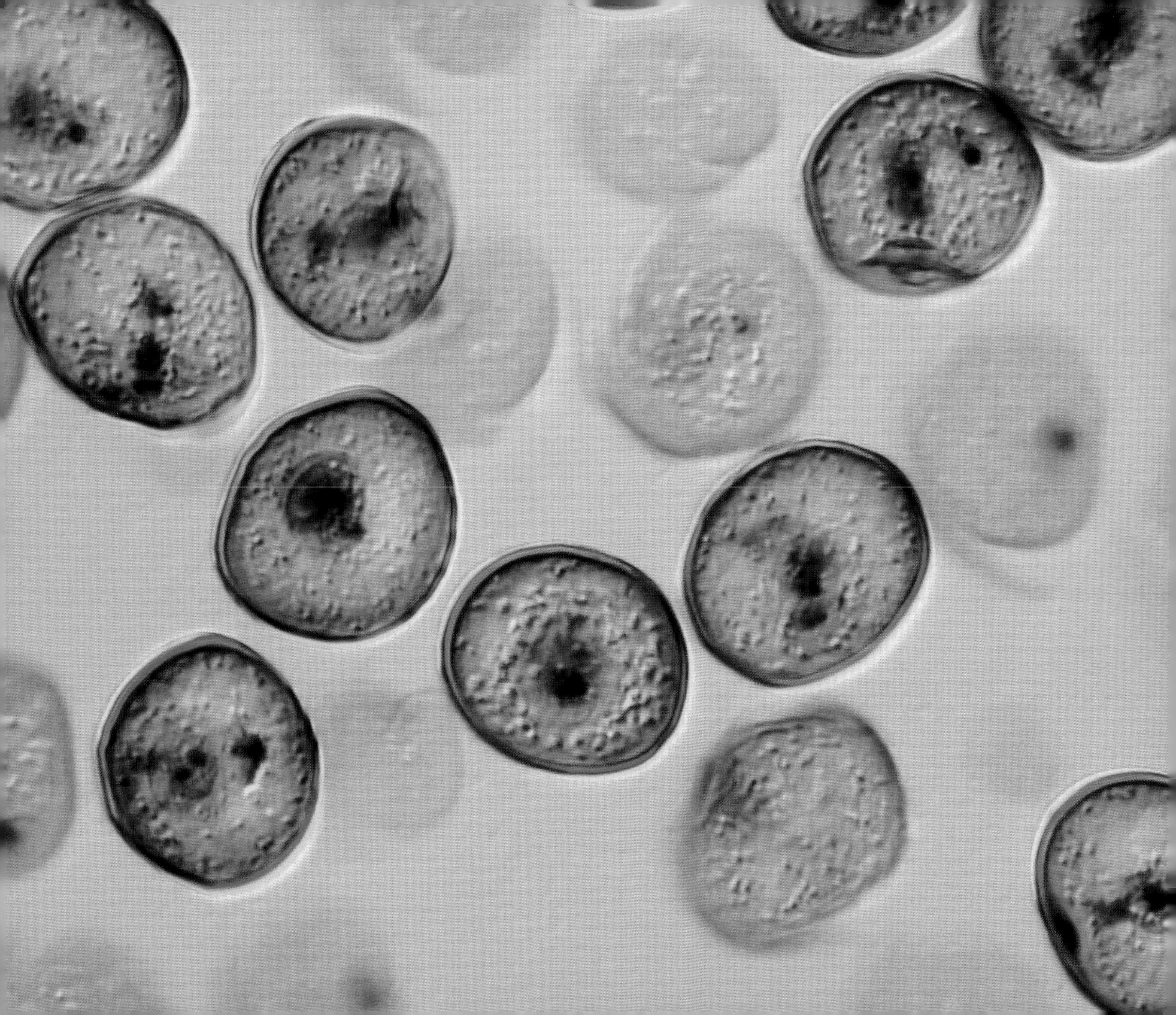

Clouds & bubbles

tell-tale trails from subatomic collisions

The 'cloud chamber' was invented by Charles Wilson at Cambridge University in 1897, as a result of his efforts to study the weather and reproduce certain cloud effects in miniature in his laboratory. A dense water vapour is formed in a test chamber. A piston is then withdrawn so that the vapour expands into a near-vacuum which has been suddenly created in the chamber. The vapour is now cool, and will condense into droplets at the slightest provocation. When a sub-atomic particle is fired through the chamber, it collides with water molecules and disturbs their electric charge ('ionises' them). Droplets immediately gather around these islands of perturbation, leaving a discernible trail in the particle's wake. In 1952, the American physicist Donald Glaser neatly inverted Wilson's concept. In his 'bubble chamber', liquid hydrogen is gently heated to the point where it is on the cusp of becoming gas, but the tight enclosure of the chamber holds it in liquid form. Then the pressure inside the chamber is eased slightly, after which subatomic particles smashing into hydrogen atoms create tiny bubbles in the liquid.

Each of these broadly equivalent techniques allows researchers to see and photograph particle tracks with conventional optical microscopes. Throughout the twentieth century, cloud and bubble chambers have revolutionised sub-atomic physics by allowing researchers to measure how magnetic fields or other influences affect particle tracks – a strong magnetic field, for instance, will curve an electron's path. We can also monitor the results of particle collisions and interactions. They may deflect or attract each other, or create showers of new particles, most of which then create their own little tracks. Some particles will escape the chamber, while others may vanish within a fraction of a second. We can study the histories of fabulously complex particle 'events', yet we never see the particles themselves. The rules of quantum physics state that we cannot measure both the momentum and position of a particle with certainty. We cannot simultaneously define its exact location in space and how fast it is travelling at that moment in time; the more we know about one value, the less sure we can be about the other. Tracks can appear beguilingly precise, but the tiny bubbles or droplets that define them can be anywhere from a hundred to a million times larger in diameter than the particles that created them. This leaves plenty of room for quantum uncertainty.

Tracks in a bubble chamber are made from strings of tiny bubbles created when particles smash through liquid hydrogen. This circular image was photographed through a microscope, and has been colour enhanced. The curved tracks are the result of magnetic fields surrounding the chamber.

Atom smashers
particle accelerators

In their efforts to study the atomic realm, Victorian experimenters heated thin metal filaments inside glass vessels (cathode ray tubes) from which most of the air had been expelled. As the filaments heated, they released streams of electrons, and magnets were used to deflect and steer the electron streams in different directions. Typically the electrons, invisible to the human eye, were guided towards a screen of zinc sulphide. When they collided with the screen, their kinetic energy was converted into visible light. Investigators then used ingenious reasoning to calculate some of the physical qualities of electrons. This is the principle by which glass television monitors still work today – and it is also the basis for the most powerful instrument on earth: the atomic particle accelerator.

Particles are generated in various ways prior to their injection into an accelerator: from a heated cathode, or perhaps by smashing a pre-existing stream of heavy particles into a target to knock off other particles. They may also emerge spontaneously from radioactive substances. The particles then travel through a vacuum down a long copper tube, the main body of the accelerator. Powerful microwave generators (called klystrons) clad the outside of the tube, creating fast-moving electromagnetic wave fronts which greatly accelerate the particles inside the tube. Electromagnets also running the length of the accelerator keep the particles confined in a narrow, controllable beam. Bubble or cloud chambers waiting at the end of the tunnel then record particle tracks (as microscopically thin trails of bubbles or condensation droplets) from which various qualities, such as mass, momentum, location or electric charge can be inferred. Modern accelerators are huge. The Stanford Linear Accelerator in California, for instance, is three kilometres long.

Circular accelerators, such as the facility at CERN, near Geneva which is 27 kilometres in diameter, propel their particles around a circular track thousands of times. At each pass, the electromagnetic fields around the giant machine are readjusted so that the particles accelerate yet further. Only when they have reached velocities close to the speed of light are they at last diverted by a shift in the surrounding control fields towards a detector. Typically, streams of billions of particles are intersected, in the expectation that just a few will collide inside a detector chamber where the results can be recorded. The engineering precisions are daunting. Several kilometres of equipment, weighing many thousands of tonnes, may be required to merge beams no wider than a human hair.

In simple terms, these are the principles by which the largest, heaviest and most expensive scientific instruments in the world probe the very smallest entities of matter.

Computer-aided design is essential in planning the huge yet extremely precise machinery of nuclear physics. This model shows a particle detector assembly at CERN, designed to filter off and analyse particle products from two separate ring-shaped accelerators.

Matter in the mirror

the discovery of antiparticles

In 1928, physicist Paul Dirac predicted that the electron should have a shadowy counterpart, an 'antiparticle' with the same mass but opposite electrical charge. In 1932, Carl Anderson observed the new particle experimentally, and it was named the 'positron'. This was the first known example of antimatter, but it soon became clear that all particles have antimatter counterparts – even, strangely enough, the neutron, which has no electrical charge for which any 'opposite' charge could exist! In 1995, scientists at CERN created an entire anti-hydrogen atom by combining an anti-proton with a positron (the normal hydrogen atom consists of one proton and one electron). But, when these anti-hydrogen atoms were produced, they survived only a few thousandths of a second before disappearing.

When matter and antimatter come into contact, they cancel each other out in a blast of pure gamma energy, the most energetic form of electromagnetic radiation. According to today's 'standard model' of physics, the Big Bang that created our universe should have created just as much antimatter as matter. So how come our universe still has so much matter left in it? Present theory suggests that if particles outnumbered antiparticles in the Big Bang by as little as one part in 100 million, then the existence of our universe could be due to the tiny fraction of particles that had no antiparticles to wipe them out. Other theories suggest that even if identical amounts of antimatter and matter were created in the Big Bang, their respective physics may have been slightly different. This difference might have favoured the survival of matter for some mysterious reason.

Low-energy positrons are routinely used in medical imaging. They are the result of the natural decay of radioactive materials. High-energy antimatter particles are only produced at a few of the world's largest particle accelerators. The current worldwide production of antimatter amounts to a few billionths of a gram per year. This is probably just as well. If lumps of antimatter the size of talcum powder grains were allowed to come into contact with matter, the explosive release of energy could destroy the laboratory conducting the research.

Are we sure there is no antimatter remaining in the universe at large? Not quite. Astronomers have discovered some evidence for antimatter, near the centre of our Milky Way, by observing gamma energies apparently created when positrons and electrons collide and annihilate.

A colour-enhanced bubble chamber photograph elegantly demonstrates the symmetrical production of matter and antimatter. Energetic gamma particle collisions with an atomic nucleus produce electrons, shown in green, and anti-electrons (positrons), highlighted in red.

Fundamental physics

searching for the simplest entities

When subatomic particles are persuaded to collide in an accelerator, their interactions can be studied. Powerful surrounding magnetic fields deflect particles to varying degrees, and the relative curves of their paths reveal their identities. If they collide with sufficient force, showers of other particles come into existence, perhaps with life spans of billionths of a second, but which quickly decay into other types of particles.

Underneath the umbrella categories of 'atoms', 'protons', 'neutrons' and 'electrons', a much more complex family of particles seems to exist. The terminology of particle physics is bewildering, but the 'standard model', developed over the last 30 years, has remained an extremely reliable tool for predicting the outcomes of many particle interactions. Matter is built from the stable (long-lasting) members of a family of particles called fermions – essentially electrons, protons, neutrons, and the yet more fundamental quarks from which protons and neutrons are built. The forces that hold subatomic components together arise from the exchange of different kinds of 'force-carrying' particles called bosons – including gluons which bind together the protons and neutrons in the cores (the nuclei) of atoms, and photons which convey electromagnetic forces. So far, there has been no experimental proof that quarks or electrons might be made from anything even smaller.

Are we finally approaching a complete description of how the world is made? Well, no. For instance, gravity is observed to work at the scale of planets and galaxies, but seems to have no observable effect on subatomic particles. How do we reconcile the conundrum of a universe so obviously shaped by gravity, yet made entirely of particles that are unaffected by it? For instance, gravity is observed to work at the scale of planets, suns and galaxies. Its influence determines the grand architecture of the universe, but its effect on individual particles cannot be measured, or even explained by the 'standard model' of particle physics. Yet the universe is made of particles. Furthermore, it is not yet understood where matter actually derives its mass – surely one of the most overt characteristics of the physical world. In 1964 Peter Higgs predicted the 'Higgs boson', which explains why particles have the mass that they do, but no one has observed one.

These problems illuminate a major philosophical issue. Every time we discover physical properties – gravity, for instance, or the masses and electric charges of particles – we need to establish how those properties are derived. In turn, the factors that create them are understood in terms of yet more underlying causes. In 1961, William Pollard, director of the Oak Ridge Institute of Nuclear Studies, summed up the daunting riddle of physics: 'Whatever it is that explains a property cannot itself possess that property … The ultimate solution of the atomic quest will be to arrive at a really elementary component of matter which accounts completely for all the diversity and behaviour of matter, yet does not itself possess any observable property whatever.' This idea goes against the grain of many physicists' hopes and beliefs, but the alternative is even stranger: that we might, in theory, carry on dissecting the subatomic world ad infinitum, without ever touching the bottom.

This stunning bubble chamber image has been colour-coded to enhance and clarify the bewildering array of particle tracks created. Bubble chamber technology is no longer the principal method of recording collision events. Computerised electronic detectors are now in standard use.

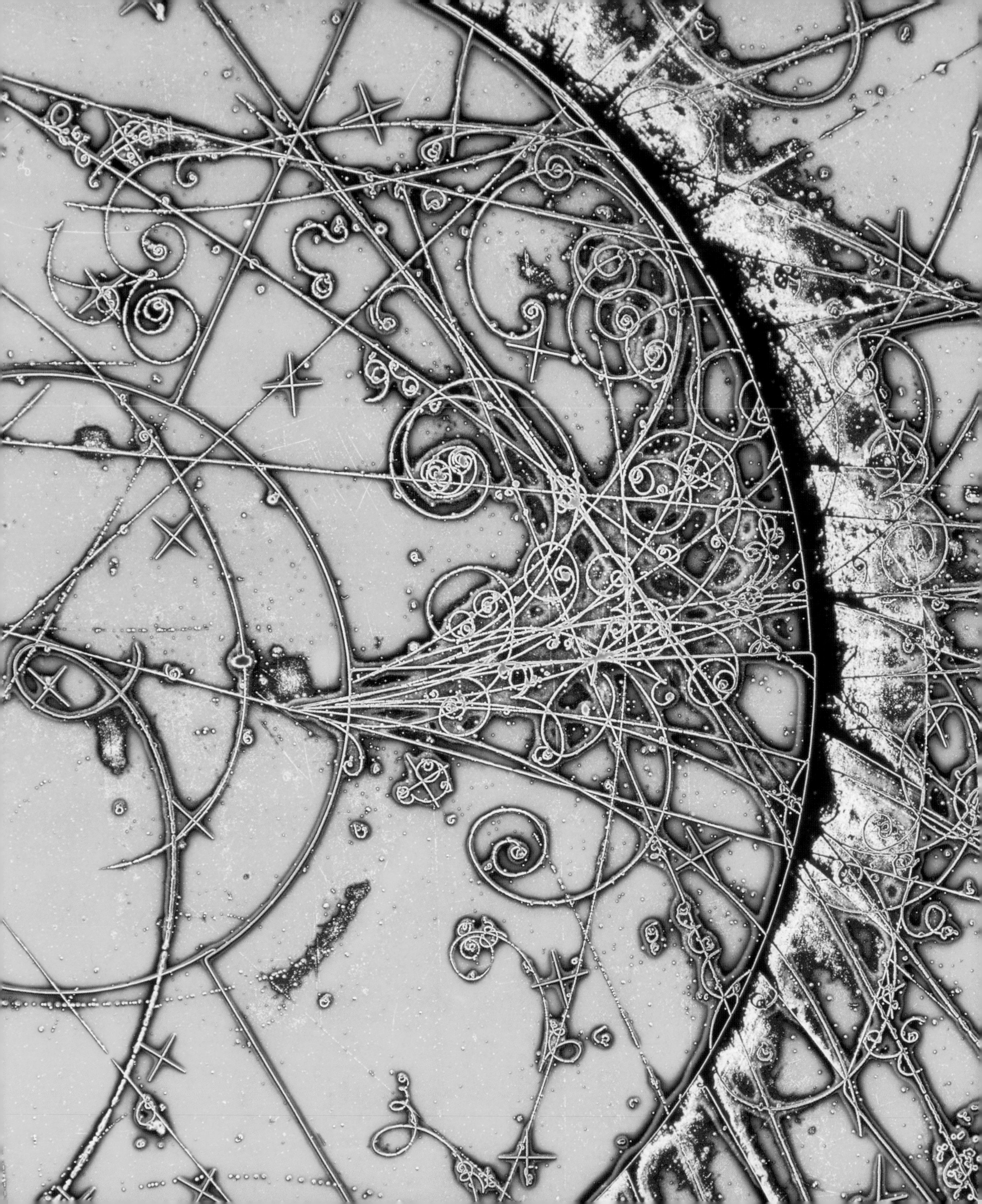

The particle zoo

mesons, fermions, bosons, leptons, muons, gluons . . .

All atoms are made up of three particles: protons and neutrons (bound together in a central nucleus) surrounded by a cloud of orbiting electrons. The question physicists were asking half a century ago was whether these were composed of even more fundamental constituents, like Russian dolls nested inside each other. They now believe they know the answer. The electron seems to be indivisible, truly 'fundamental', while protons and neutrons are now thought to be made from a family of yet smaller particles, called quarks. But these building blocks do not fully explain how an atom actually works. For instance, positive protons and negative electrons cancel each other's electrical charge in an atom, creating a stable, balanced arrangement, but what about the neutrons, which have no electric charge? What holds them in place? In the 1930s, physicists realised that some additional force, not an electric charge, has to glue the neutrons together. Japanese physicist Hideki Yukawa predicted the meson, a particle that leaps between protons and neutrons, carrying 'exchange forces' to bind the neutrons into the nucleus. The old 'solar system in miniature' model of the atom was no longer sufficient to explain how matter works. Experiments over the next half-century showed that perhaps as many as 200 particles needed to be taken into account to describe the particle interactions observed in experiments and plug gaps in the mathematical theories. There are pions, kaons, etas, muons, taus, electron neutrinos, muon neutrinos, and so forth, divided among familes of baryons, hadrons, fermions, bosons, leptons…

One great difficulty is that many particles seem to have incredibly brief lives when observed in an experiment. They blink out of existence in tiny fractions of a second, yet the world around us obviously relies on particles that persist over immense spans of time. Every single atom in our bodies, whether carbon, oxygen, hydrogen or nitrogen, dates from at least the time of the solar system's formation 5 billion years ago, and most of them may well have been drifting through space, as part of the debris from long-dead earlier generations of stars, before being gathered up by our Sun and its planets. We decay and die, but our atoms move on. So long as an atom is not radioactive, caught up in a nuclear explosion, or sucked into the fiery fusion core of a sun, it is indestructible. Atoms may gain or lose electrons, becoming positively or negatively charged, but that condition is reversible. The 'carbonness' of a carbon atom is permanent.

And think of all those photons of electromagnetic energy (gamma, X-ray, ultraviolet, visible, infrared, microwave and radio) hurtling across the universe for billions of years, until at last they impact with our telescopes. Obviously they have persisted for vast spans of time. Yet within the powerful accelerators used to conduct physics research, particles of bewildering fragility and short duration are regularly observed. Some of them vanish in a few billionths of a second. Of course, accelerators create a special set of conditions that are only rarely (if ever) found in nature, especially in the routine high-energy experiments designed to smash heavy particles together at near-light speeds. Often, these experiments artificially create the conditions just after the Big Bang. But that is very different from telling us about how matter behaves today.

A computer display of subatomic particle tracks, spraying out from a high-energy collision in the Large Hadron Collider at CERN in Switzerland. 'Hadron' is the collective term for all particles that can be bound together inside atoms by the 'strong nuclear force'.

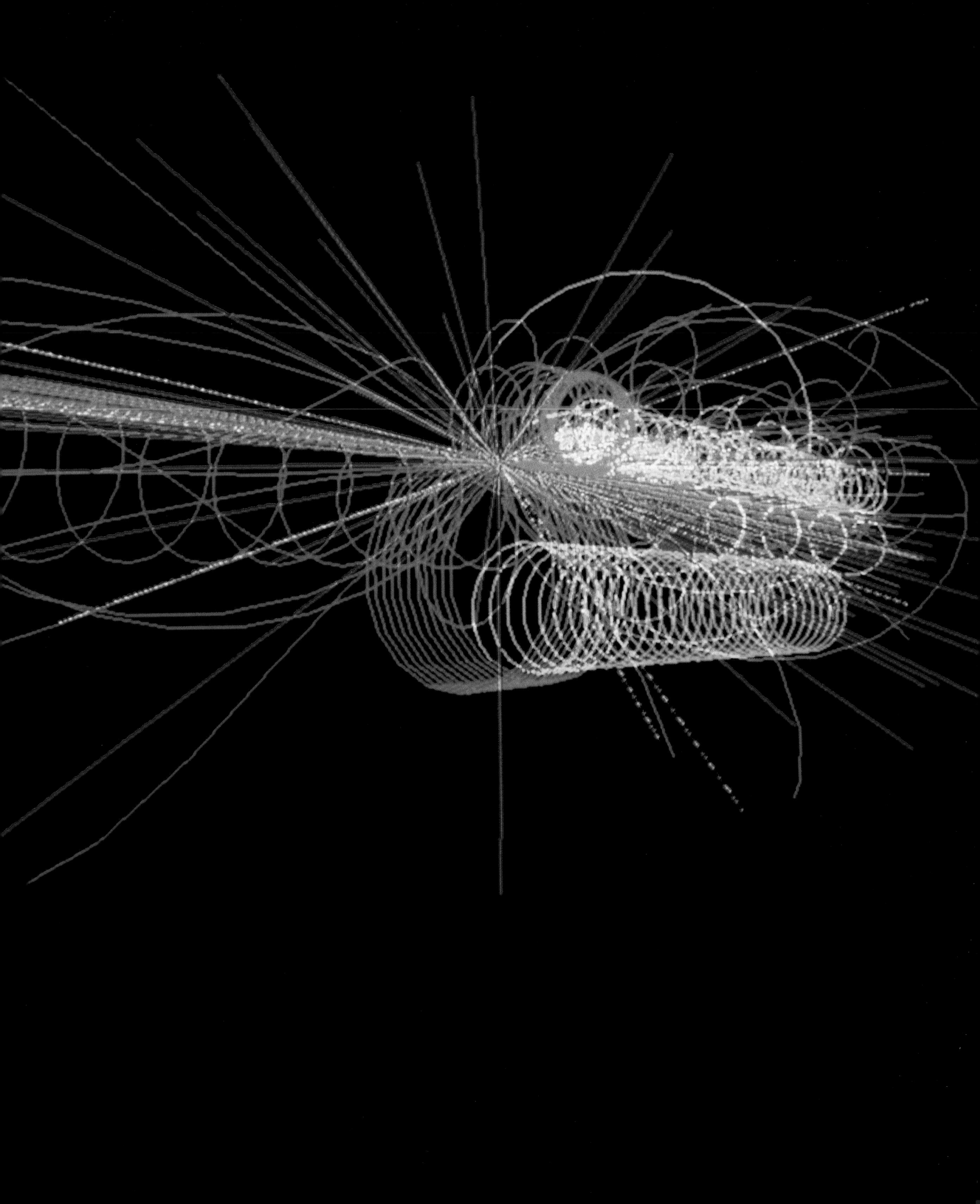

Electronic physicists

smart computers
interpolate particle trails

There have always been drawbacks associated with cloud and bubble chamber images of particle tracks and collisions. Firstly, as accelerators grow more powerful, and tracks ever more numerous and complex, it becomes harder to derive precise data. A chamber can be so heavily criss-crossed by tracks that it becomes hard to tell one from the other merely by visual inspection of the photographic plates. Secondly, and much more significantly, information about very small-scale particle events may be compromised, because incidental collisions with the vapours or gases inside the chambers inevitably affect the particles under study, even as they create the tracks we see. One of the perpetual frustrations of atomic physics is that our physical instruments, built from matter, can only detect the properties of particles by interacting with them in some way. According to the rule known as 'quantum uncertainty', any observation of a particle made with a physical device must alter its behaviour. On the other hand, if we do not measure it with an instrument, we discover nothing about it.

Modern accelerators dispense with any intervening medium, be it vapour or liquid. This at least minimises some (but by no means all) of the ambiguities. Collisions take place in a pure vacuum. Electronic detectors then register particles emerging from these collisions, which may fly off in any direction. Complex and extremely expensive arrays of detectors must completely surround the collision zone so as not to miss important clues. A wide variety of detectors has been designed to respond to particular types of particle, but in principle they are not entirely unlike the electronic chips inside digital cameras: they generate tiny electrical signals when incoming particles dislodge electrons inside them. These are amplified and sent to computers for analysis.

The benefit of these techniques is that collisions occur in pristine perfection; the downside, inevitably, is that the collisions themselves cannot be observed – only their products, striking detectors in the walls of the test chamber. Computers as powerful and sophisticated as we are capable of building must judge what types of particle, and what shapes of collision, are most likely to have produced those 'hits'. They learn from repeated experience (by cross-referencing many similar experiments) how to distinguish different events. They then make graphic recreations of the detector arrays, showing where hits have been registered, and how energetic they were. Finally, they simulate the collisions, making three-dimensional pictures of the particle tracks – essentially 'predicting' unseen happenings after the fact. But the computers can only work with the theories of subatomic physics that have already been programmed into them by human scientists. To some extent at least, they may be showing us what we already expect to see.

This display from the Collider Detector Facility at Fermilab, near Chicago, shows the result of a collision between a proton and an anti-proton. The spokes represent different zones in the detector. Red and yellow bars signify particles generated in the collision, along with their energies.

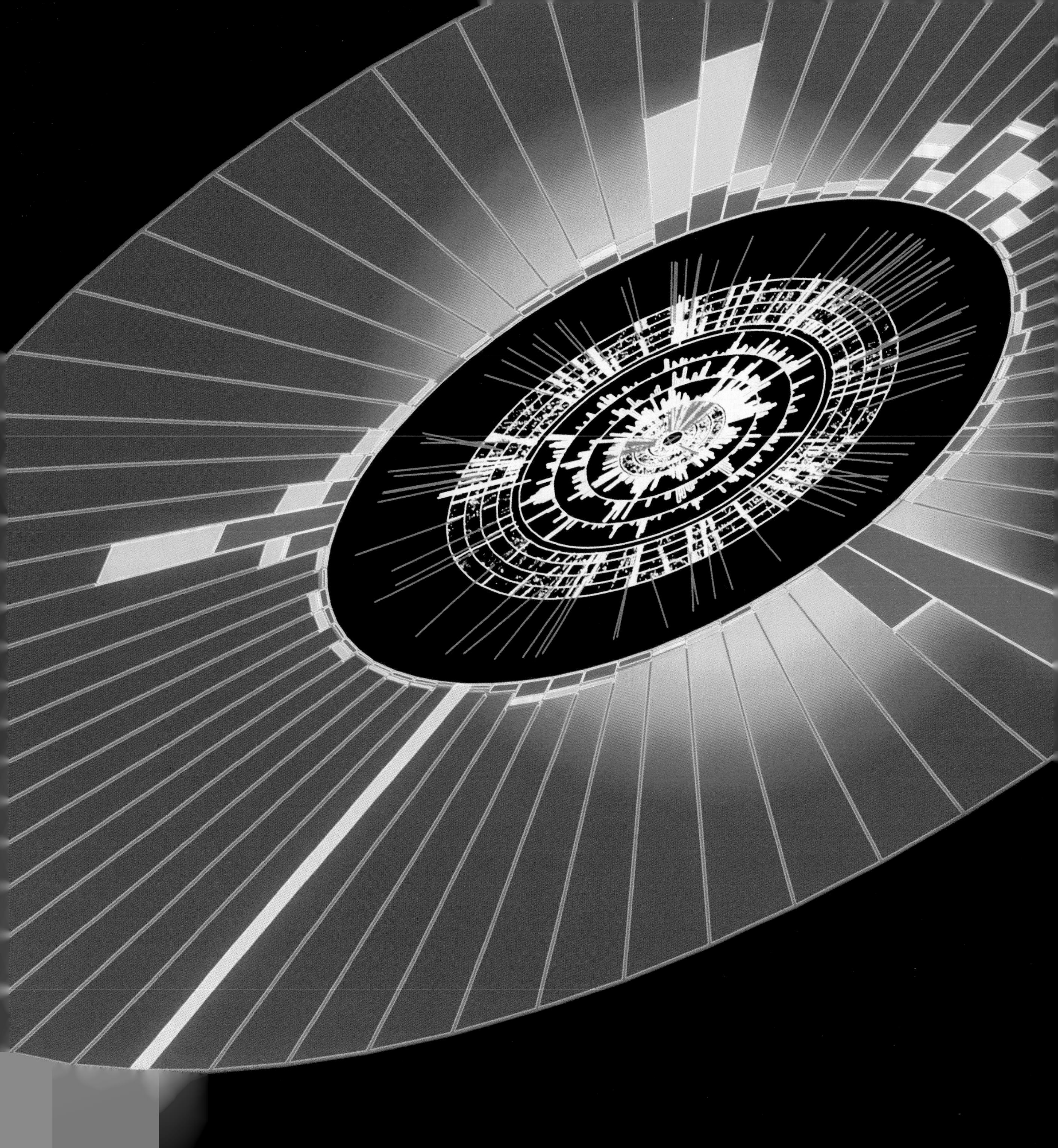

Recreating the cosmic dawn

searching for the origins of matter

In June 2003, physicists at the Brookhaven National Laboratory's Relativistic Heavy Ion Collider (RHIC) announced they had recreated the extremely hot, dense, high-energy conditions that might have existed in the first millionth of a second after the Big Bang that gave rise to our universe, some 13.7 billion years ago. Beams of heavy gold nuclei were accelerated to near-light speeds and then smashed together. The collisions were so violent, the debris they produced briefly reached temperatures of one trillion degrees centigrade (more than 300 million times the surface temperature of the Sun). The protons and neutrons in the gold nuclei disintegrated, releasing the quarks from which they were made and the gluons that bound them together.

The strange material fleetingly created in the experiments was labelled a 'quark-gluon plasma'. A plasma is a high-temperature gas consisting of subatomic particles rather than whole atoms. The quark-gluon combination revealed what might briefly have existed before the universe began coalescing, within the first millionth of a second after the Big Bang, into ordinary atomic matter.

However, some physicists cautioned that the Brookhaven data did not necessarily show individual quarks and gluons freed from their bonds, so much as streams of other particles created and thrown out by the massively energetic experiment. The liberation of quarks and gluons was inferred from theoretical explanations rather than positively observed. Theory suggests that it is not possible to observe a free-flying quark in the laboratory, because the collision energies required to free them (by, for instance, smashing protons apart) are so great that new particles are created which are themselves made out of quarks. Nevertheless, the Brookhaven scientists are confident that their investigations into the earliest fractions of a second of the universe's history are sound. This image shows a computer interpretation of the particle trails that were actually recorded in the experiment.

An impressive spray of particle tracks flies off during quark-gluon plasma research. Heavy gold nuclei are smashed into each other at near-light speeds to create densities of matter 20 times greater than in lead or plutonium, and far hotter than the heart of the Sun.

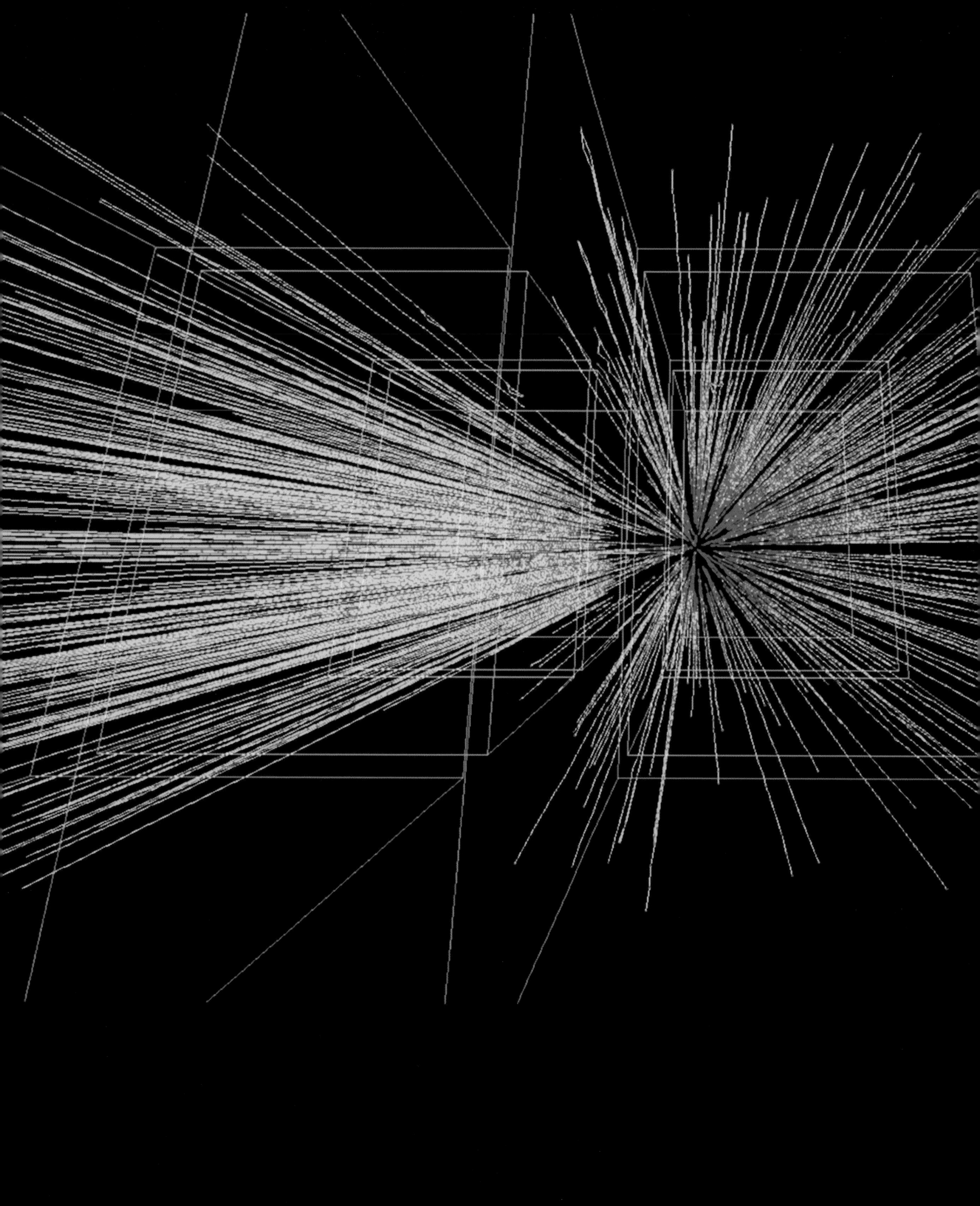

Crystals of creation

X-ray diffraction of molecular arrays

Imagine a chandelier, with hundreds of glittering cut-glass facets and baubles. The room is dark, the light switch is off, but a narrowly focused torch beam is shining from one side of the room and through the chandelier. By the time the torchlight reaches the opposite wall, it has been bounced around to such an extent by the chandelier's many surfaces, it emerges split into countless directions. The resulting pattern of light on the far wall seems to bear no relation to the physical outline of the chandelier, or indeed to the original neatly focused torch beam. It reveals, instead, how the chandelier diffracts the beam as it passes through. If dozens of chandeliers are probed in this way, each one being brought into the room and suspended from the same ceiling hook, and the torch in each instance being directed in the same way, then the different patterns on the wall can be correlated with the individual chandeliers that created them. Eventually, by rotating a chandelier in increments and analysing the shifting patterns of light on the wall, an expert observer might be able to create a model of the chandelier's general structure: the spatial relationships between its hanging glass prisms and globes.

This analogy is a good way of visualising how X-rays are used to explore the structures of chemical compounds – but it is just an analogy. Unlike chandeliers, it is impossible to investigate objects as tiny as atoms or molecules using ordinary light, because optical wavelengths are far too large. X-ray wavelengths, on the other hand, are extremely small. When they hit an atom, the surrounding electrons diffract the beam, which then forms a characteristic pattern (an X-ray diffraction) as it exits. Heavier atoms with many electrons scatter X-rays to a greater extent than lighter ones with fewer electrons. This has allowed scientists to explore the structures of complex molecules to a high level of precision.

Fluids and gases are not so amenable to diffraction experiments because they are in constant motion. Samples need to be crystallised into forms that hold their shape rigidly. While X-ray data from a single molecule would be too ambiguous for sensible analysis, crystals make good subjects because they are built from 'unit cells', arrays in which molecular patterns repeat themselves many times, amplifying the diffraction effects. But sample crystals need to be extremely tidy. Any impurities distort the results, and even gravity corrupts the geometry of a crystal – imagine a chandelier suspended too low from the ceiling so that it partially drags on the floor. Efforts are currently under way to grow perfect crystals in weightless conditions aboard the International Space Station.

The X-ray diffraction pattern from rubisco, a compound widely found in photosynthesising plants (below). In the larger image, the many facets of a testosterone (male sex hormone) crystal break up the alignment of polarised light waves.

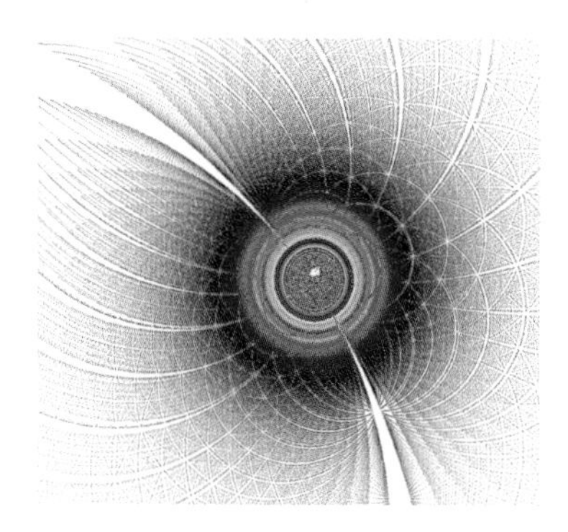

Virtual chemistry
building molecular models with computers

Before the advent of computers, the structures of complex molecules had to be inferred by applying the known rules of chemistry. Chemical deconstruction of a molecule would reveal the ratios of atomic elements within it, and then polystyrene balls and bits of wire, representing the constituent elements and their chemical bondings, would be assembled as much by brilliant guesswork as anything else. In 1957, James Watson and Francis Crick constructed the first logically accurate representation of the DNA double helix using cardboard cut-out shapes, plastic balls and metal rods held together with screw clamps on a steel pole.

Allied to X-ray crystallography, computers can now speed up the process of calculating what a complex molecule might look like. 'Virtual' molecules can be combined to output a new molecule, and then the computer predictions can be matched with combinatory experiments using real molecules. Piece by piece, the complexities of biochemistry are being revealed. But the proteins of life are dauntingly complicated. Take ribosomes, for instance. They manufacture all the proteins in a body, but the way they do so is not yet understood. It was only as recently as 1999 that a crystallised ribosome was finally modelled accurately, after many years' research. An average protein contains 300 amino acids and, in theory, there are millions of ways in which it might be constructed without breaking the rules of chemistry. Every time a protein is studied, its exact structure needs to be discerned because its shape vastly influences how it functions.

Even the most basic organic compounds can confound us. In the 1960s, for instance, the sedative drug thalidomide was thought to be well understood. It was a simple compound of carbon, oxygen and nitrogen. Yet, the difference between a left-hand or right-hand twist in an otherwise identical chemical array turned out to be the difference between medical success and human tragedy. The left-handed version was entirely safe, but the right-handed molecule, when taken by pregnant women, caused severe abnormalities in their unborn children. The power of computers now enables a vast array of virtual molecular experiments to be made prior to any medical applications in the real world, but the immense complexities of proteins will take many years to unravel yet. For instance, proteins are amazingly dynamic: within fractions of a second of being created, they fold themselves into intricate shapes. We need to understand not just those final shapes but how and why the folding occurs. Dozens of diseases owe their origins to misfolded proteins.

A virtual computer model of botulinum, a nerve toxin commonly known as Botox. It is used to paralyse muscles in the face, either for cosmetic purposes or to treat distressing squints and other tics. The model can be manipulated on screen, and even viewed with 3D glasses.

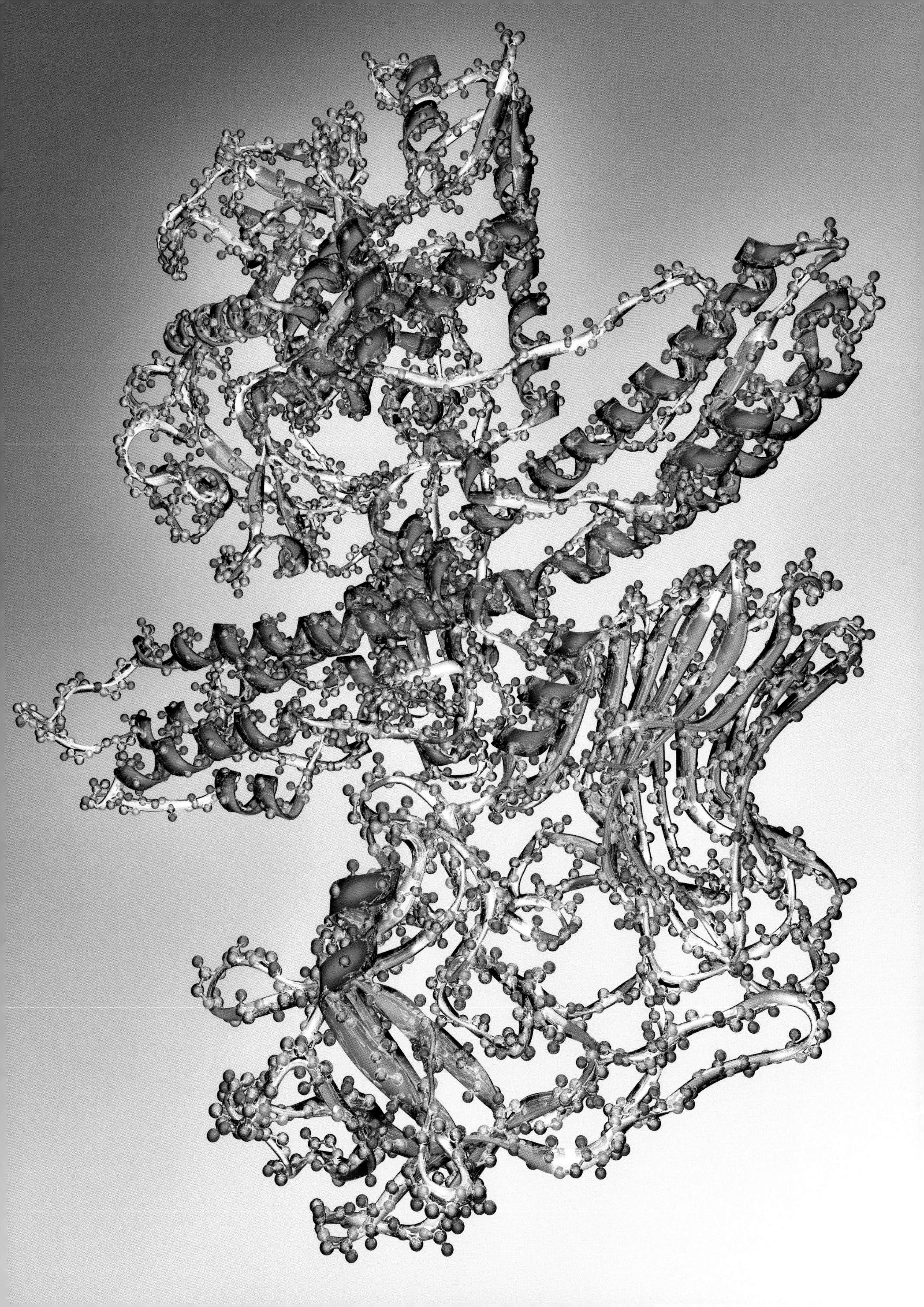

The limits of vision

scanning tunnelling microscopy of atoms

What is the smallest thing that we can see? In 1981, Gerd Binnig and Heinrich Rohrer of the IBM Zurich Research Laboratory invented the scanning tunnelling electron microscope (STEM). A super-fine tungsten probe – its tip may contain just a single atom – is moved across the surface of a sample. The movements are controlled by nothing so crude as gears and wheels, but by extremely small 'piezoelectric' quartz crystals that change their shape very slightly when an electric current is applied to them. As the crystals distort, they move the probe. When the tungsten probe scans its target, a measurable charge flows between the tungsten atoms and those in the sample, because electrons jump (or 'tunnel') across the gap via a process called quantum tunnelling. The piezoelectric crystals also act as a feedback device, measuring how much counter-force is required to keep the probe at a constant distance from its target. Atoms can be magnified 100 million times and visualised in the form of contour maps. The main disadvantage of the STEM is that it can only be used with conducting and semi-conducting materials, including metals, because materials that do not conduct electricity would not allow the tunnelling effect to occur. A wide range of related instruments have emerged in recent years to solve this problem: for instance, the atomic force microscope (AFM) uses a diamond tip, and does not require a transfer of charge between the tip and its target. Consequently, AFM machines are suited to examining non-conductive biological samples, such as viruses and proteins. The diamond probe registers superfine mechanical displacements as it is dragged across a surface

Of course, STEM machines do not produce anything that we might sensibly recognise as a 'picture' of an atom; rather, the machine generates images on a screen in response to particular electric and magnetic forces, and creates a map of atomic locations. Nevertheless, STEM is an impressive technology that will no doubt become ever more refined over coming decades.

A colour-coded STEM image of uranium atoms inside a microscopic crystal. Each spot represents an atom. The distance between the centres of adjacent atoms is approximately one 10-millionth of a millimetre. The STEM has magnified the array of atoms by a factor of 22 million.

The duality of matter

are particles objects or waves?

The wave-like properties of light were demonstrated by the English physician and physicist Thomas Young in the early nineteenth century. He shone light through two narrow, closely spaced slits in a metal sheet and saw, on a screen placed behind it, how a pattern of wave interference fringes was created. In the twentieth century, physicists wondered whether particles of matter could also behave like waves. When the 'double-slit' experiment was repeated using electrons, an amazing thing happened. A focused beam of single electrons seemed to pass through both slits at once, so that wave-like interference was observed on the other side of the screen. When one slit was blocked off, the interference pattern vanished. How did the electrons 'know' in advance whether or not they were going to encounter one or two slits? When electrons were fired just one at a time, at intervals of many seconds, the same results were observed: wave interference patterns when firing through two slits, and a clean vertical line of bullet-like 'hits' (no interference pattern) when only one slit was open.

With single electrons fired at such long intervals apart, there is no possibility of them interfering with each other mid-stream. They must, essentially, be interfering with *themselves.* But how can an electron decide in advance whether or not it is going to pass, wave-like, through two slits at once, or else behave like a discrete particle and travel cleanly through just one slit? Another variation: if two slits are left open, and electron detectors are placed just behind them, then once again the interference patterns vanish, and the electrons revert to particle-like behaviour, definitely deciding to go through one slit or the other, but not both at once. Again, how can they know in advance that detectors are waiting on the other side of the slits?

'Particle-wave duality' presents us with a disturbing conundrum. Our very act of observing a particle seems to affect, in advance, what it will do, and how it will appear to us. Most theoreticians are happy to accept the topsy-turvy abstractions of the 'quantum' world, where particles can be both point-like bullets of matter or vaguely dispersed waves, can go backwards in time, and be in several places at once. Thinking about what it all means, however, is much more of a challenge.

Waves in a water tank convey some of the qualities of the 'double-slit' quantum wave experiment. Wave fronts on the left of the picture encounter a screen with two small holes. Two sets of circular waves emerge on the right, mingling to create interference patterns.

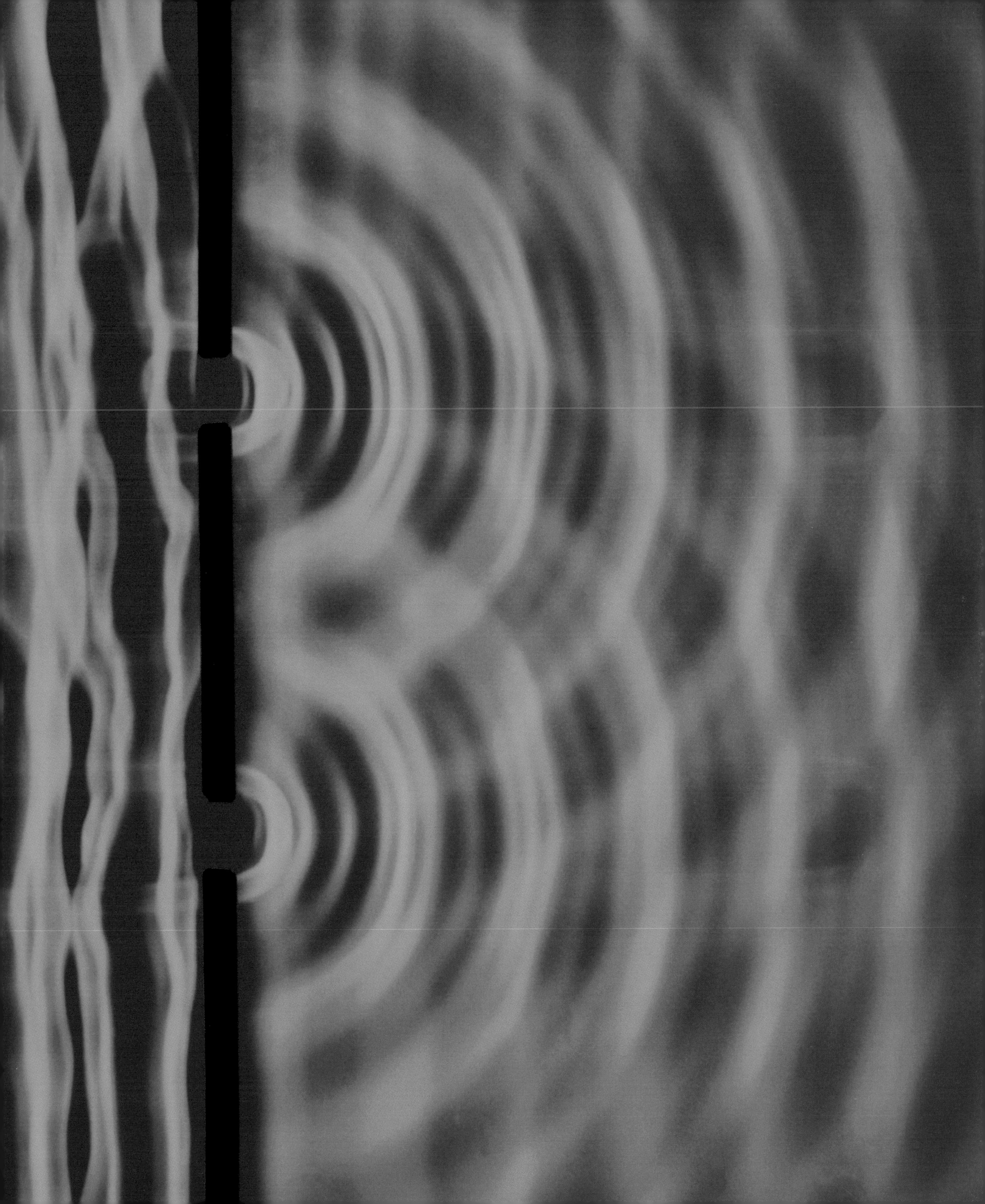

Subatomic shimmer

images of quantum waves

In 1993, Don Eigler and his colleagues at the IBM Research Center in California released the first images of a 'quantum corral'. This is perhaps the most extraordinary picture yet produced by any human agency, because it visualises some of the eerie qualities at the heart of quantum physics. A century ago, atoms were thought to consist of billiard ball-like protons and neutrons in a central cluster (the nucleus), orbited by electrons, like tiny moons held captive by a planet. Today, quantum theory tells us that all subatomic particles exist in a shimmering cloud of possible locations and states – until a measurement is made, at which point some specific qualities may be observed: in particular, location or momentum. But the more accurately we measure one quality, the less we know about the other. Depending on the experiments we choose to make, we can find a particle at a particular place at a particular time, or we can know how fast it is moving at that particular time, but we can never pin down both measurements at once. Quantum theory treats particles as ghostly waves of probability rather than as discrete objects located definitely in both time and space. Eigler's team set out to capture an image of electrons as waves.

A scanning tunnelling electron microscope (STEM) was used to deposit a tight circle of atoms on a smooth base, using an extremely fine electrically charged metal tip. The team chose cobalt atoms on copper for their initial experiments. When the cobalt atoms were pushed into a sufficiently tight circle, the electron wave forms associated with each atom interfered with one another, building up or cancelling each other out. They generated 'standing waves' inside the circle of atoms that the STEM tip could measure with electric and magnetic forces. The quantum corral vividly showed electrons refusing to be observed as being anywhere in particular. The more finely the instruments probe, the more elusive subatomic particles become. This STEM image is not a 'picture' of electron waves, because such a thing is impossible. It is a computer model extrapolated from electric and magnetic measurements, with the colours tuned to make it more easily readable. The whole experiment can be compared with Braille. The STEM is a sensitive fingertip, allowing our minds to read the subatomic words and phrases to which our eyes must forever remain blind.

The image on the right shows the quantum corral produced by Don Eigler and his team. In another experiment (below) the STEM has arranged the surrounding corral of atoms into an ellipse, so that the electron waves shimmer around two focus points.

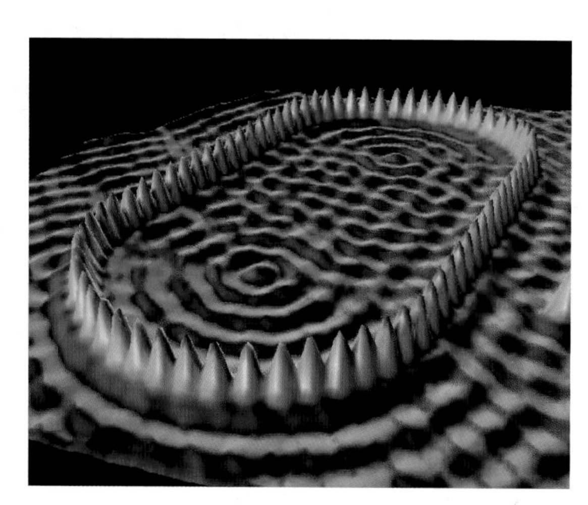

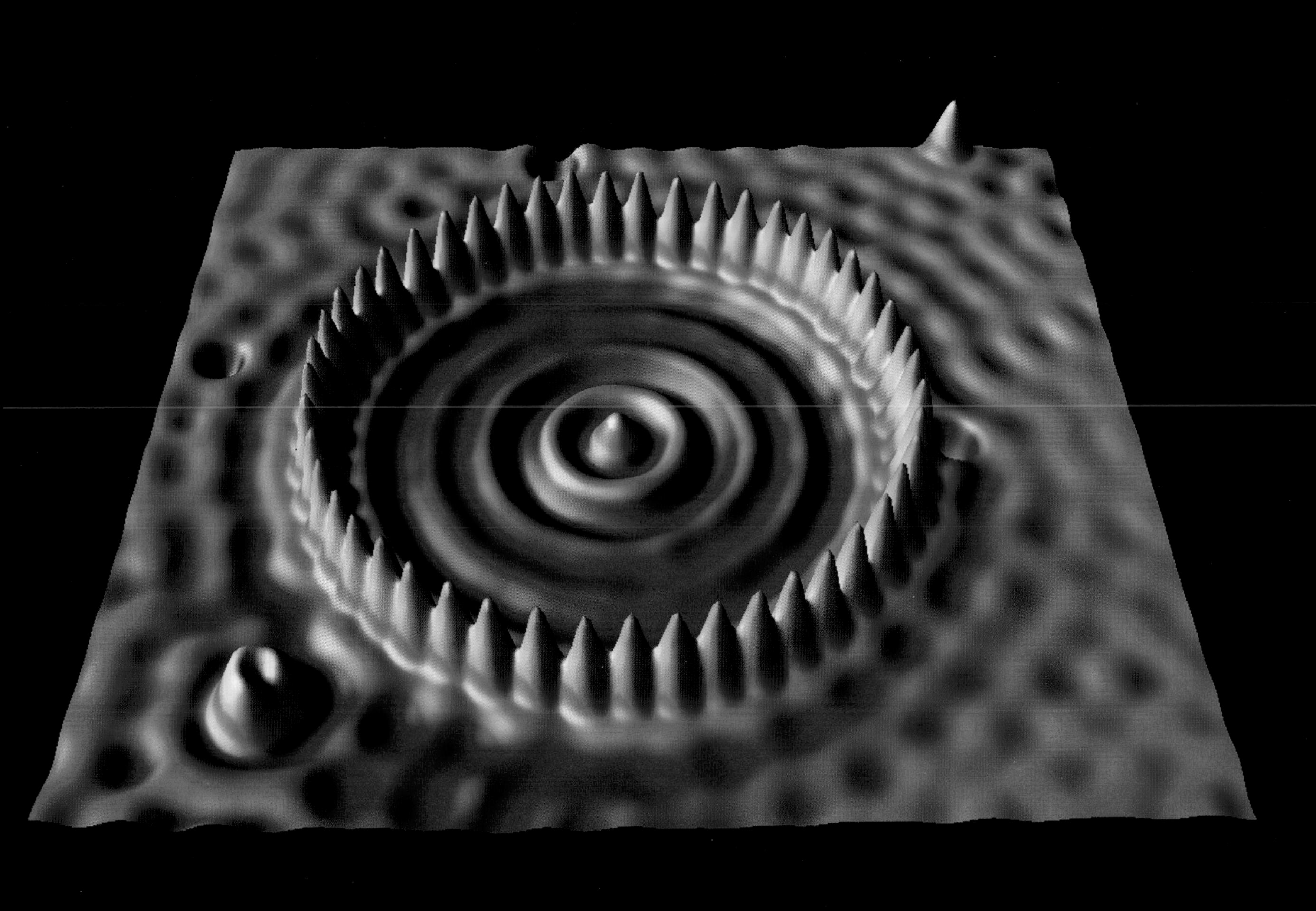

Circuits in action
the hidden life of a microchip

In 1959, two separate inventors, Jack Kilby and Robert Noyce, conceived what we now call semiconductor 'microchip' technology. Kilby used polished crystals of germanium as his base material and Noyce used silicon, but the principles each adopted were the same. Transistors, resistors, and all the other components of an electronic circuit were traditionally manufactured as separate items, then wired onto a printed circuit board held together with molten gobbets of solder. It was a costly, labour-intensive process, and the clumsy wiring set limits on how tightly components could be packed together. That in turn affected the size and cost of electronic products. Kilby and Noyce pioneered a way of printing components in miniature onto polished crystal wafers.

The process works rather like a photographic enlarger in reverse. A big, detailed image of the required circuitry is reduced via lenses to the size of a stamp, and beamed onto a light-sensitive coating on the surface of the crystal. Chemical baths etch away specific areas in the coating. Subsequent coatings and etchings pile on more and more components. Although the production line is expensive to build, it eventually churns out many thousands of chips at a very small unit cost. Today, 125 million transistors can be squeezed together onto a single chip the size of a penny.

In the mid-1960s, Gordon Moore, an engineer at Hewlett-Packard, calculated that the number of transistors on a microchip was doubling every 18 months, while the cost to consumers was halving in that same time. Now referred to as Moore's Law, his calculation has held up. Many of the electronic toys children buy with their pocket money have more computing power than an IBM mainframe of the 1960s that filled a large room. A typical car has a better computer than the Apollo rocket that travelled to the moon in 1969 – and at less than one millionth of the price.

But the microchip may soon be reaching its limits. The wavelength of visible light is only so small, and the enlarger-in-reverse printing process can only scale things down so far. Photo etching has been replaced by ultraviolet techniques, and even X-ray lithography, using (nearly) the smallest possible wavelengths of light to define the tiniest possible details of a circuit. The problem is that individual components become so small, they interfere with their neighbours at the quantum level. The clearly defined flow of electrons along particular wires breaks down into something less distinct.

A new computer revolution is required. Nanotechnology may hold the key, building up from the molecular level with unparalleled degrees of accuracy so that quantum interferences are kept to an absolute minimum. But it is just as likely that designers will learn to harness the quantum effects, transforming not just the stuff from which computers are made, but also the very nature of the software they use. Say goodbye to the simple binary language of 'on' or 'off' pulses that has defined computing for 60 years. Welcome to the world of 'both on and off at the same time'.

A microscope image of a fairly ordinary silicon chip shows it to be an impressive achievement, but it could be reaching the point of obsolescence. Some are concerned that the 'information economy' will experience a severe downturn unless we invent new and better technology.

The nano-world

reshaping matter at the smallest scales

The images derived from a scanning tunnelling electron microscope, or any of the vast range of related instruments available today, are not really snapshots of atoms or quantum waves, but computer extrapolations several times removed from the original objects. The contours are graphic representations of electric forces, and colour intensities can represent magnetic effects. The atomic realm does not really display straightforward characteristics, yet the human imagination rebels against this truth. We cannot help but imagine that if we could peer a little closer, or make a STEM tip even finer, we might at last be able to 'see what an atom really looks like'. We have to accept that the building blocks of matter have no form that could possibly be interpreted by our human senses. Nevertheless, we *can* manipulate them.

In 1959, physicist Richard Feynman made a prediction which, at the time, sounded purely theoretical: 'The principles of physics do not speak against the possibility of manoeuvring things atom by atom. In principle, it should be possible to synthesise any chemical substance. How? Put the atoms where the chemist says, and so you make the substance.' Three decades later, nanotechnology pioneer Eric Drexler popularised the idea of achieving 'thorough control of the structure of matter at the molecular level. It entails the ability to build molecular systems with atom-by-atom precision, yielding a variety of nano-machines.' In 1990, Don Eigler and Erhard Schweizer of the IBM Almaden Research Center in California spelt out the letters 'IBM' with 35 xenon atoms arranged on a nickel surface, using the tip of a tunnelling electron microscope to manoeuvre individual atoms.

Manufacturing – say, of a fine clockwork watch – usually involves taking large chunks of metal and whittling them down into the required cogwheels and springs. It is an inefficient procedure, because most of the metal ends up on the workshop floor as discarded shavings. In addition, huge amounts of energy are needed to extract pure metal from crude ores in the first place, and then to melt the metal into bars, plates or rods that can be yet more finely cut up and shaped by the watchmakers. Nanotechnology turns all this processing on its head, building upwards from the smallest scales by molecular accretion. Biology functions like this all the time. We are now entering an age in which some of the distinctions between biology and engineering are starting to blur; we will *grow* our machines, rather than build them.

An electron microscope image of a 'micro-machine', a miniaturised electric motor etched onto a wafer of silicon in such a way that the various components can move freely. The image below shows atoms arranged into the Chinese symbols denoting the word 'atom'.

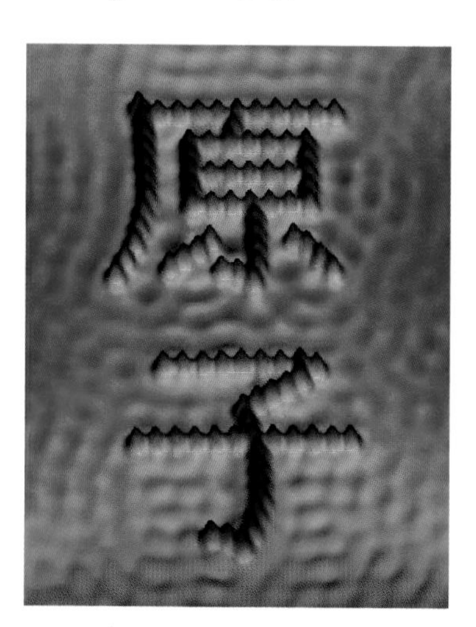

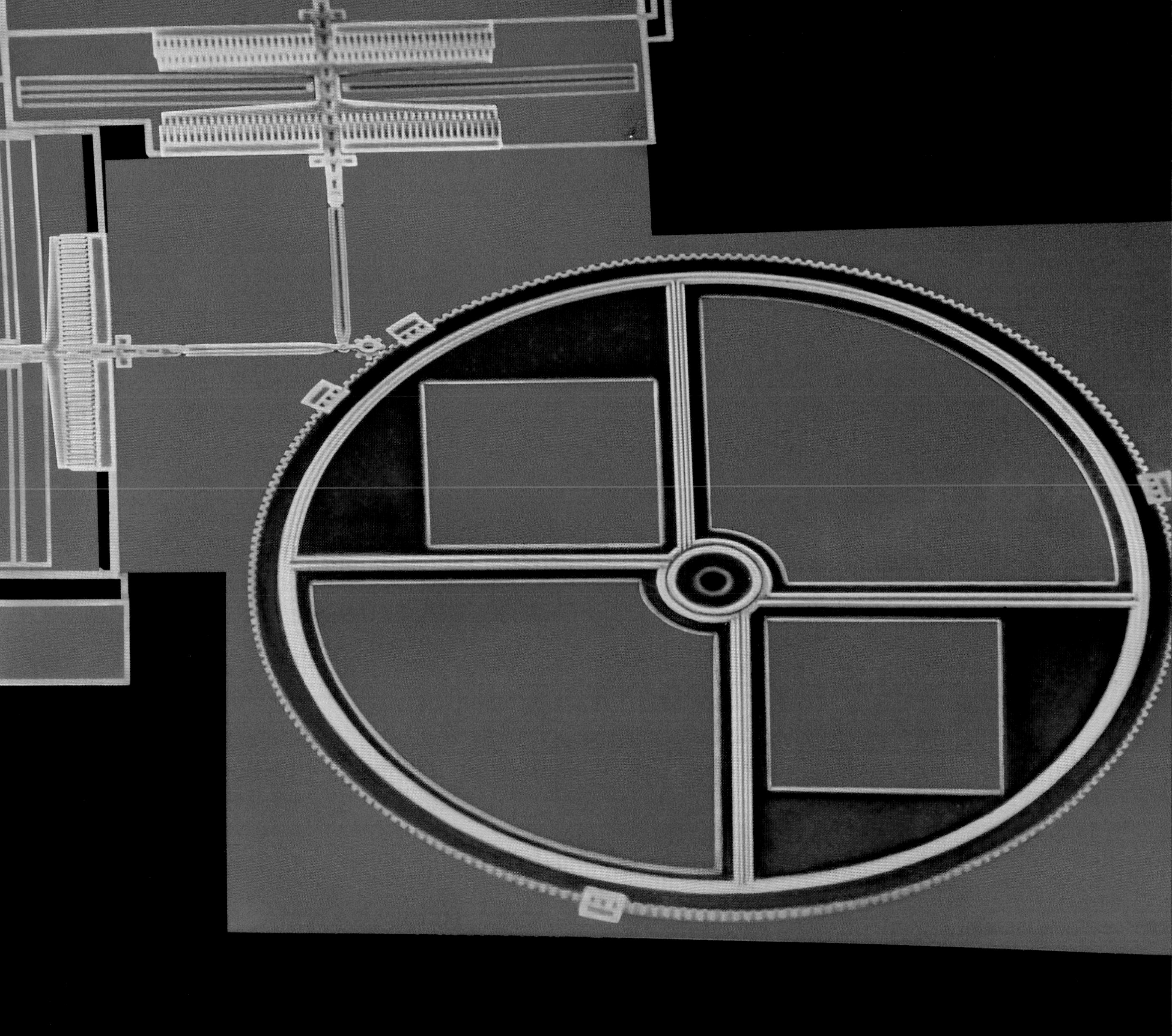

2

Bugs in the machine

That insignificant speck of dust on the table could turn out to be a monstrous predator with hairy legs, sharp mandibles and an armour-plated body. Or maybe it is a speck of dust after all? It takes a powerful kind of electronic eye to tell for sure. Electron microscopes can enlarge the smallest bedbug thousands of times and reveal it in sharpest detail. Our carpets, curtains, mattresses, even our clothes provide a living for hordes of creatures. Our bodies, too, reveal amazing secrets when individual hairs, flecks of skin and blood cells come under the electron microscope. And there is even a chance that fossilised bugs in a chunk of rock could prove the existence of life on Mars – just so long as we know how to read the electronic images properly. Perhaps those fossils are not what they seem?

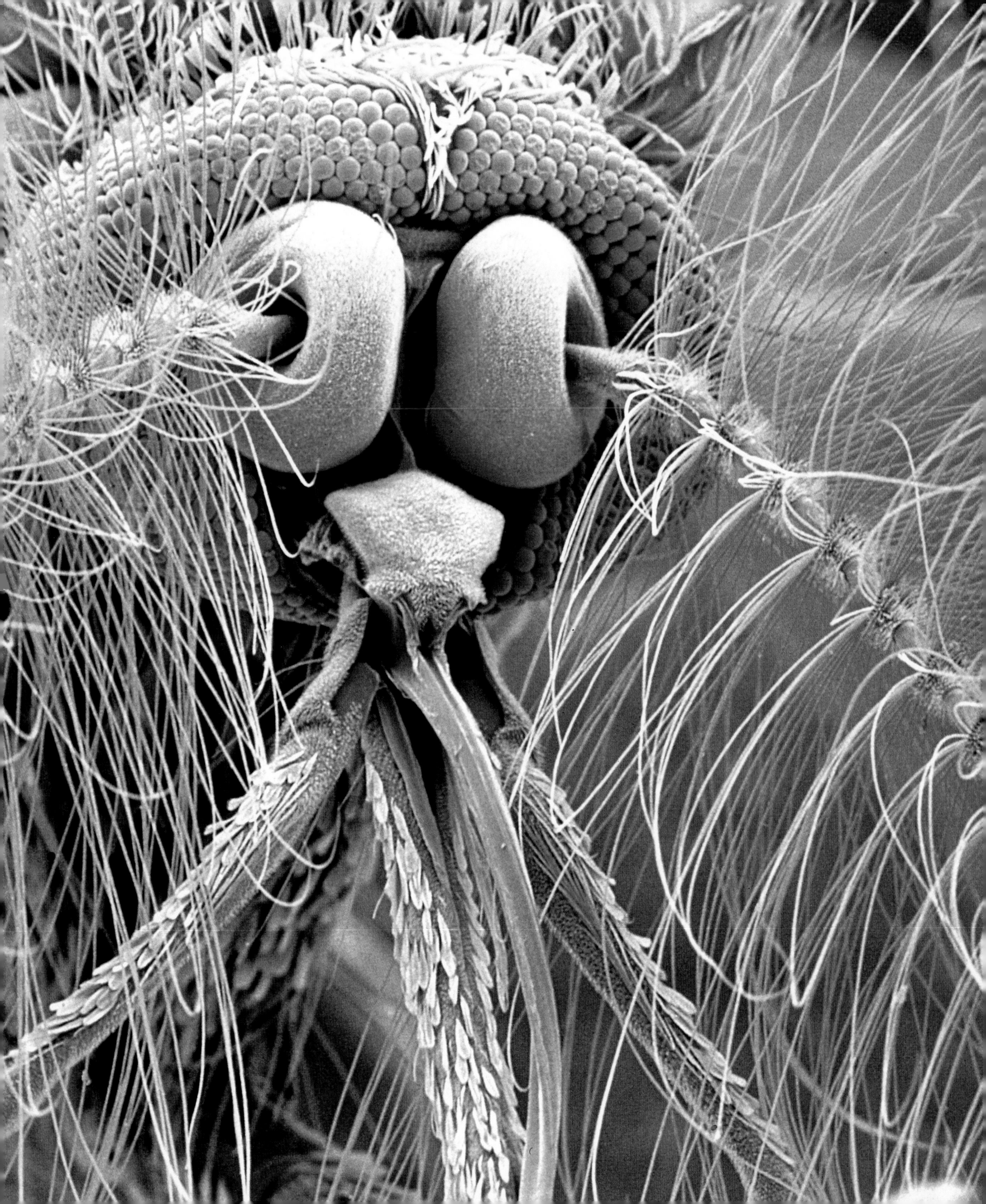

The borderlines of life
are viruses truly alive?

Anyone suffering from a common cold virus infection feels as if they are under attack by a 'bug', but in the scientific community, no one is quite sure if a virus is truly alive.

Bacteria (proper 'bugs') are recognisably biological cells that have walls and internal structures. They feed, excrete and replicate. Under a microscope, or swarming on the surface of a petri dish, visible to the naked eye, they are tangibly 'there' as living things. Viruses, on the other hand, are little more than strands of DNA or RNA molecules covered by a protein coating. They are a thousand times smaller than bacteria. While bacteria can reproduce independently, viruses can do so only by hijacking the cells they infect. The protein shell protecting a virus's DNA is covered with spiky protrusions, which insert genetic material into a host cell. The virus's DNA takes control and forces the cell to use up all of its resources to manufacture yet more viruses. The weakened cell eventually bursts like a balloon and is destroyed; the replicated viruses then attach themselves to new, unaffected host cells, and the viral infection continues.

Living things usually do more than just reproduce. Above all, they eat to gain energy. Bacteria have evolved an incredible variety of lifestyles, from ingesting each other or invading larger organisms (sometimes beneficially) to processing minerals and gases in the Earth's crust. They are undoubtedly alive. Viruses, on the other hand, have no internal metabolism and no need for food. They are just chemical replicating machines.

Viruses may not be alive, but they are very much a part of life on Earth, and part of the human experience, from chickenpox and measles to mumps and polio. They infect animals, plants and even bacteria. We are in a constant war with the Ebola virus, the West Nile virus, and influenza – the 'common cold' in its deadliest form. Influenza killed millions of people around the world in 1918. Today, AIDS, the illness that can result from an HIV infection, is claiming 3 million victims a year. The total number of deaths since the epidemic first became apparent in the 1980s is difficult to calculate, but is at least 21 million.

Creationists who claim that biology is the product of deliberate design, rather than Darwinian evolution, might find it hard to fit viruses into their world view. If they are the product of divine inspiration, they constitute a mean-minded addition to the world's catalogue of biochemical wonders; however, considered dispassionately in evolutionary terms, they are an amazing success story. We are learning to harness their invasive skills to insert beneficial genetic material into human cells to cure diseases and inherited ailments, so viruses may one day save as many lives as they destroy. They will become the subjects of deliberate design, but not in the way that creationists might have imagined.

Preceding pages
A mosquito's head, eyes and antennae are captured in this stunning electron microscope scan. The carefully composed and artificially coloured image seems to bridge the divide between science and art.

Facing page
In a transmission electron microscope, beams of electrons are fired through a biological sample to produce an image. This is a single Ebola virus, the agent of a particularly vicious and frequently fatal type of fever. In this image, the virus has been magnified by a factor of 64,000.

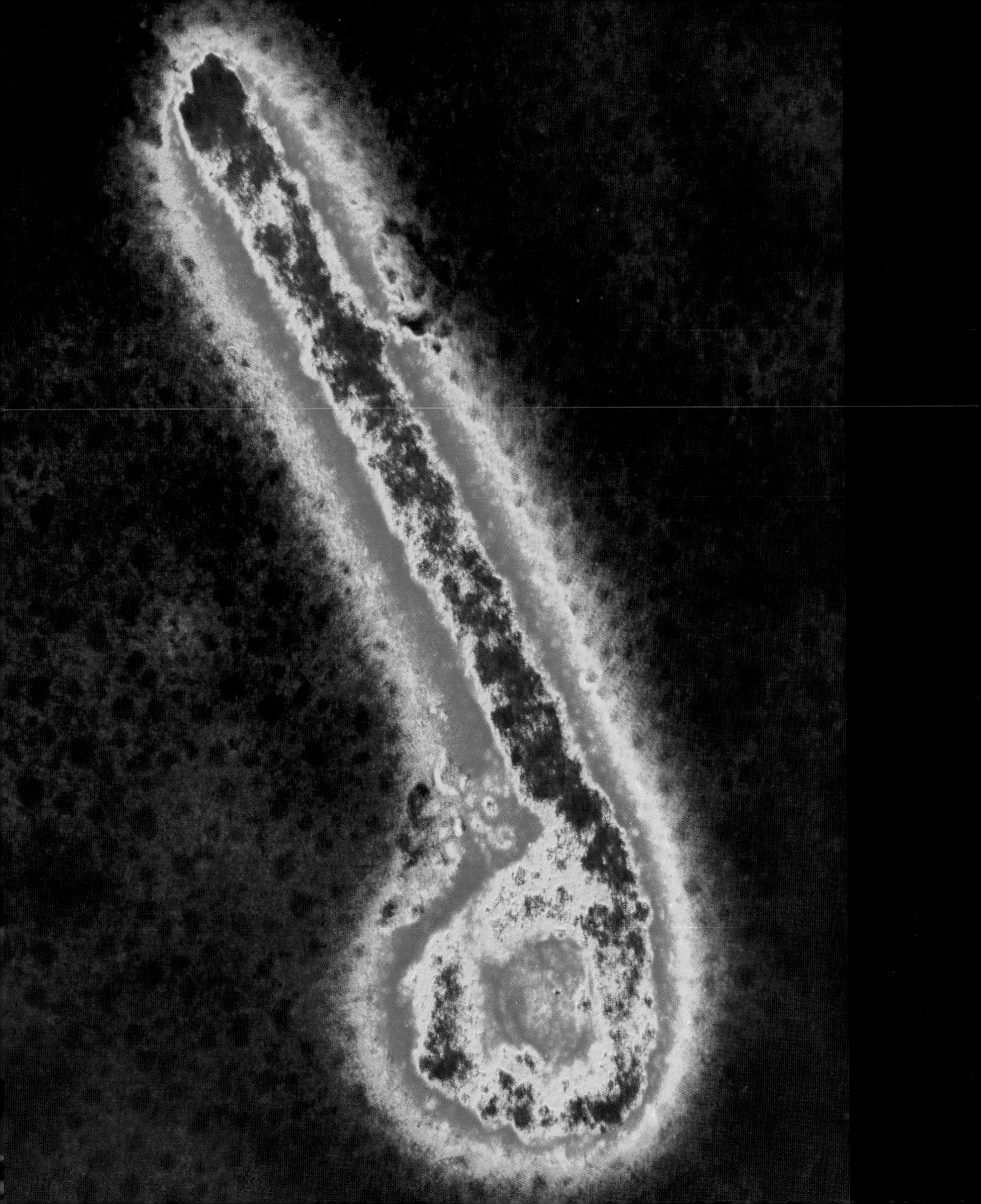

Cells & bacteria

biological organisms at microscopic scales

A conventional optical light microscope can magnify an object up to 1,000 times, and resolve details 0.0002 millimetres apart. Anything smaller than this cannot be resolved because the wavelengths of optical light are too large. The transmission electron microscope (TEM) was developed by Max Knoll and Ernst Ruska in Germany in 1931 in an attempt to break the magnification barrier imposed by ordinary light wavelengths. High-energy electrons (far smaller than the wavelengths of optical light) are transmitted through a very thin specimen or an ultra-thin section of a sample. Where the electrons pass through to strike a detector, the image is bright, and where they are scattered away, the image is correspondingly dark. Sections and thin specimens can be stained in various ways to increase electron scattering and contribute to image contrast.

Until recently, electron microscopes could not be used to look at 'wet' samples, because the electron beam that creates the image needed to travel through a complete vacuum. Samples would literally decompress, and all their water would boil away. Since just about everything in biology consists mainly of water, this was quite a problem for scientists wishing to look at delicate cells and bacteria in their natural state. When bacteria (for instance) are examined, they are first prepared in a dense solution, which is then spun in a centrifuge so that the bacterial cells end up as a dense layer at the bottom of the container. A sample of this is smeared on a thin transparent holder and allowed to dry out. Then it can be placed into the microscope.

But the discernible membranes, cell walls and other structures of biology yield little information unless we are able to witness how they work. Biology is a complex set of processes, not merely a collection of physical components. New systems – 'environmental' microscopes – allow living cells and bacteria to be studied without the need for a vacuum inside the sample chambers.

A cluster of salmonella bacteria during division – samples from a large family of bacteria that infect the digestive system and cause food poisoning (gastroenteritis). The cells are rod-shaped, and sprout long, hair-like 'flagellae' which help the bacteria to move around.

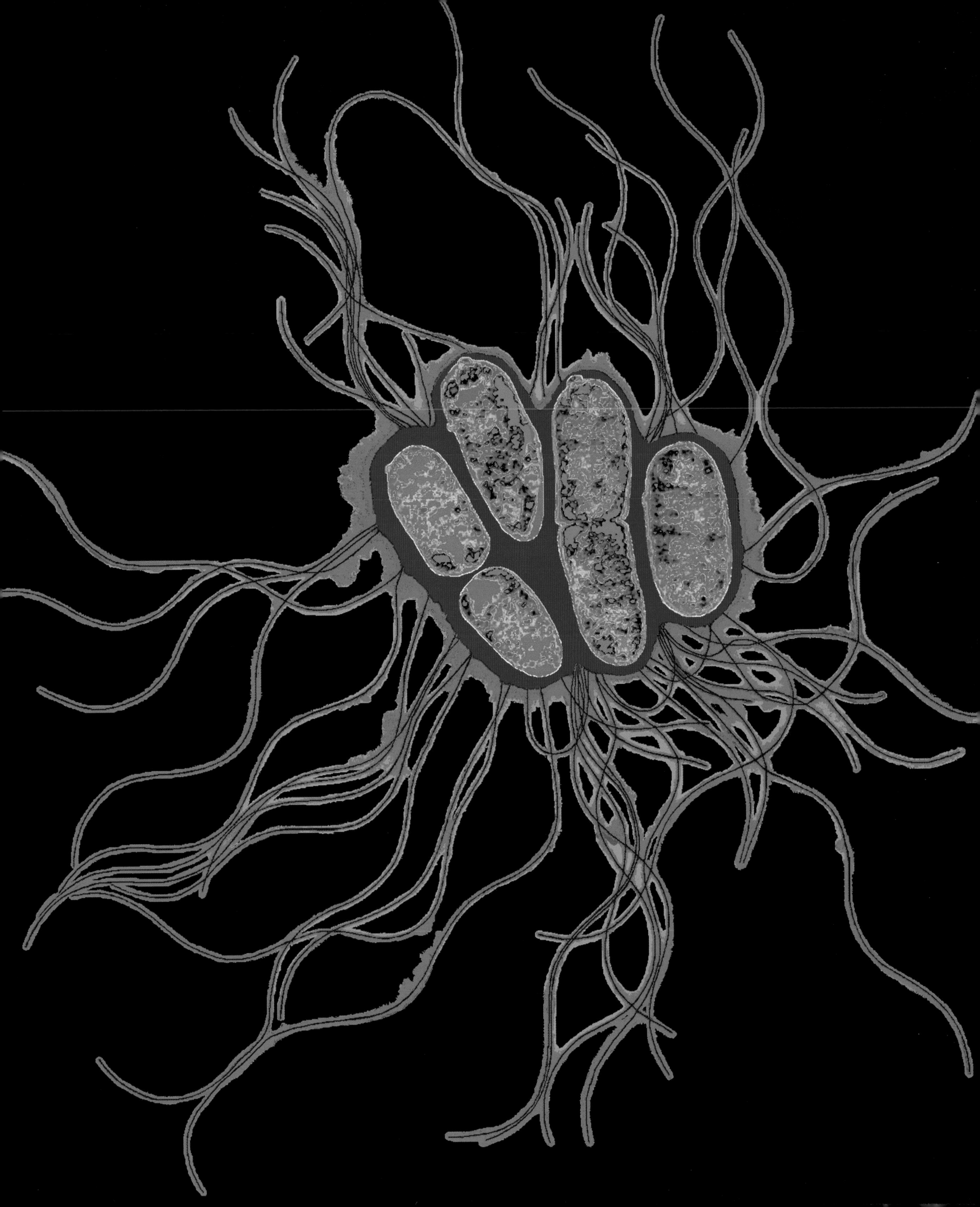

The organic glow
fluorescence microscopy

Virtually all materials interact with light in some way. They absorb, reflect or refract light that hits them. When white daylight shines on a red berry, pigments in the berry's skin absorb the green, yellow and blue wavelengths and reflect red. Our eyes therefore only see the red light coming from the berry. Fluorescent materials perform a slightly different trick. They take in light of one colour and send it back out as another.

Many substances exhibit fluorescence: phosphorus, quinine sulphate solutions and paraffin oil, for instance. In these substances, electrons in their chemical compounds are boosted into higher orbits by incoming light photons – but some of the energy exchange is dissipated as heat, and the electrons release lower-energy photons when they fall back to their original orbits. If a particular molecule absorbs blue light, it may fluoresce lower-energy green light in response. Alternatively, a molecule that absorbs green may give off yellow or red light (but never blue). Optical light is not sufficiently energetic to cause visible fluorescence, as the incoming light tends to swamp the far weaker outgoing colours. Ultraviolet wavelengths, on the other hand, are more energetic than all the optical colours, and create vigorous fluorescence – yet the ultraviolet itself cannot be seen by the human eye. Ultraviolet is often nicknamed 'black light' for this reason.

Anyone who has caught the beam of a 'black light' lamp in a nightclub will have noticed how specks of dandruff on the shoulder of a jacket will suddenly advertise themselves with a prominent blue glow. Fluorescent organic compounds in the skin cells react to the ultraviolet. In broad daylight the effect is swamped, but in the dark of a disco, where much of the optical light is dimmed, the effect can easily be seen.

This phenomenon is widely exploited in microscopy. Under a normal optical microscope, biological specimens may look surprisingly uninteresting. Transparent cells, for instance, will not always reveal their finer structures. Coloured dyes that attach themselves to different types of organic compound can highlight features more clearly, but one of the best techniques for very small-scale imaging is ultraviolet fluorescence microscopy. Special dyes called fluorochromes react to a nearby ultraviolet light source by giving out optical light that is clearly visible through the microscope (all other sources of optical light are blocked). The fluorochromes bind chemically with specific target compounds in a sample, and fluoresce at predetermined optical colours. Using several fluorochromes at once, researchers can literally colour-code structures inside cells, sometimes as small as 50 molecules across. The results can be profoundly beautiful.

A beautiful fluorescent light micrograph looks like a galactic nebula or some other cosmic wonder. In fact it shows melanoma (skin cancer) invading healthy tissue. Orange-glowing fluorescent dyes have attached themselves to proteins particular to the cancerous cells.

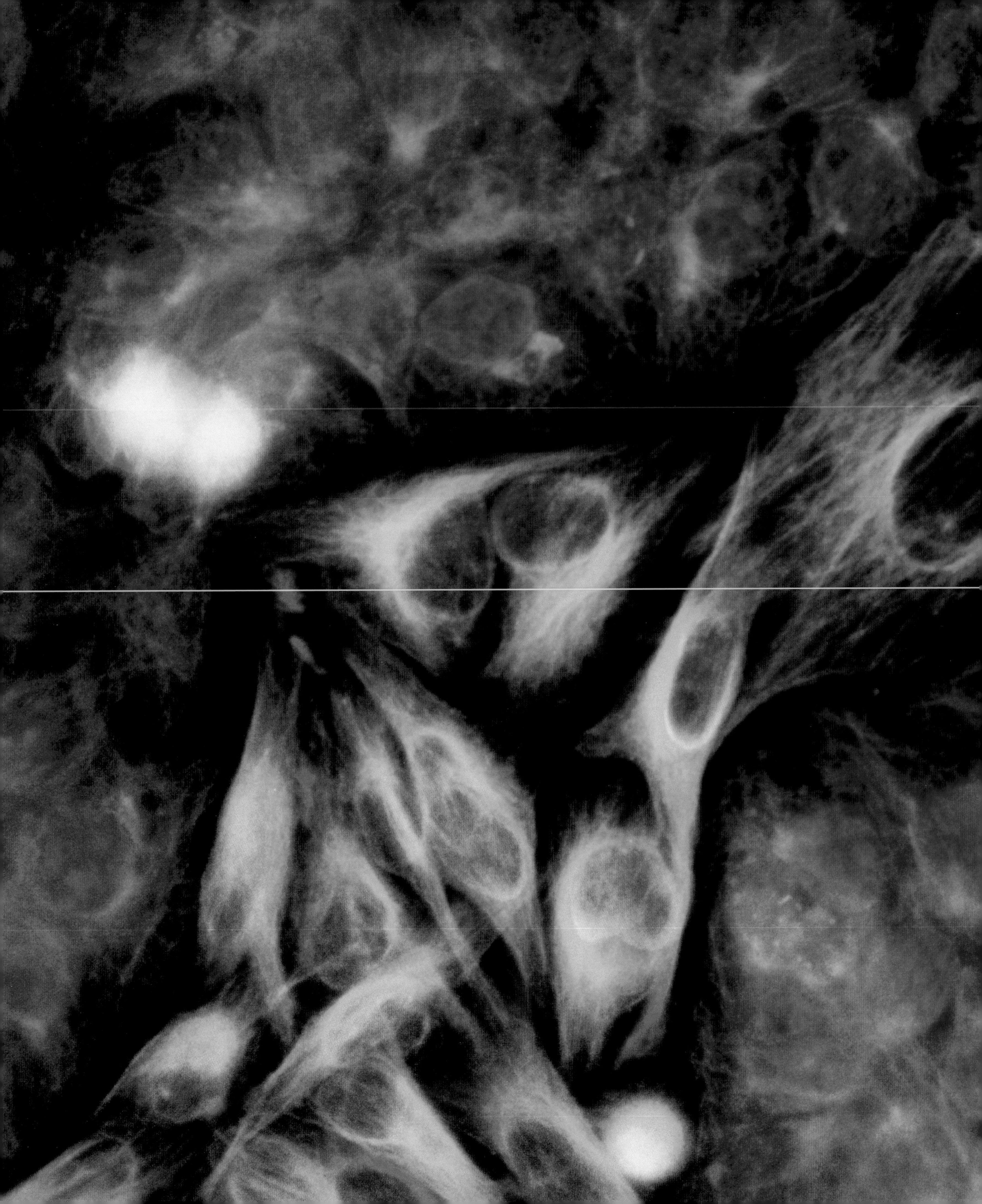

The greenness of greenery

turning sunlight into life

Plants, algae and certain bacteria perform an everyday miracle, turning sunlight energy into chemical food energy using photosynthesis. Specifically, plants absorb carbon dioxide from the air, and water from the ground, to create sugar glucose at the rate of millions of molecules every second. Plants use much of this glucose, a carbohydrate, as an energy source to build leaves, flowers, fruits and seeds. They also convert it into cellulose: the tough, indigestible structural material that gives rigidity to their cell walls. Most plants produce more glucose than they use, however, and store it as starch and other forms of carbohydrate. This is just as well, since the primary energy source for humans and other animals are those very same carbohydrates, derived either from eating plants or from eating other animals that eat plants. Each year, photosynthesising organisms produce about 170 billion tons of carbohydrates: 30 tons for every single person on the planet.

The waste product of this process is the very oxygen in the air that we breathe. Human civilisations are also dependent on the ancient products of photosynthesis. The fossil fuels that supply most of our industrial energy are composed of hydrocarbons, the remains of organisms that relied on photosynthesis millions of years ago. Virtually all life on Earth, either directly or indirectly, depends on photosynthesis.

The chemistry and physics of photosynthesis are subtle and extremely complicated. Light energy causes electrons in chlorophyll and other light-trapping molecules to boost into higher-energy orbits. When the electrons fall back, as they must, their excess energy is distributed in a fantastically complex array of electrochemical transport chains, to fuel synthesis reactions. Red and blue wavelengths of light are the most effective in photosynthesis because they have exactly the right amount of power to energise chlorophyll electrons. There are a few exceptions, but most plants reflect unwanted green light straight back out into the world – which is why plants appear green.

Four billion years ago, the Earth's atmosphere consisted largely of carbon dioxide. The oceans supported a great mass of blue-green algae that absorbed carbon dioxide using photosynthetic reactions. They excreted oxygen, which was too chemically reactive in its raw state for their delicate structures to absorb. Eventually they fell victim to their own success: by pumping out such vast quantities of oxygen, they altered the atmosphere so drastically that they killed themselves off.

This electron microscope image shows a chloroplast, the specialised cell that performs photosynthesis. It contains chlorophyll and other enzymes which cannot be artificially created or simulated by any human technology. Decades of experimentation have given us only a partial understanding of this process. We cannot turn sunlight into food the way that plants can.

An electron microscope scan of a Christmas rose leaf, cut in cross-section. The waxy upper surface of the leaf is at the top. Numerous vertical cells inside the leaf house chloroplasts, the globular structures which contain the chemical systems of photosynthesis.

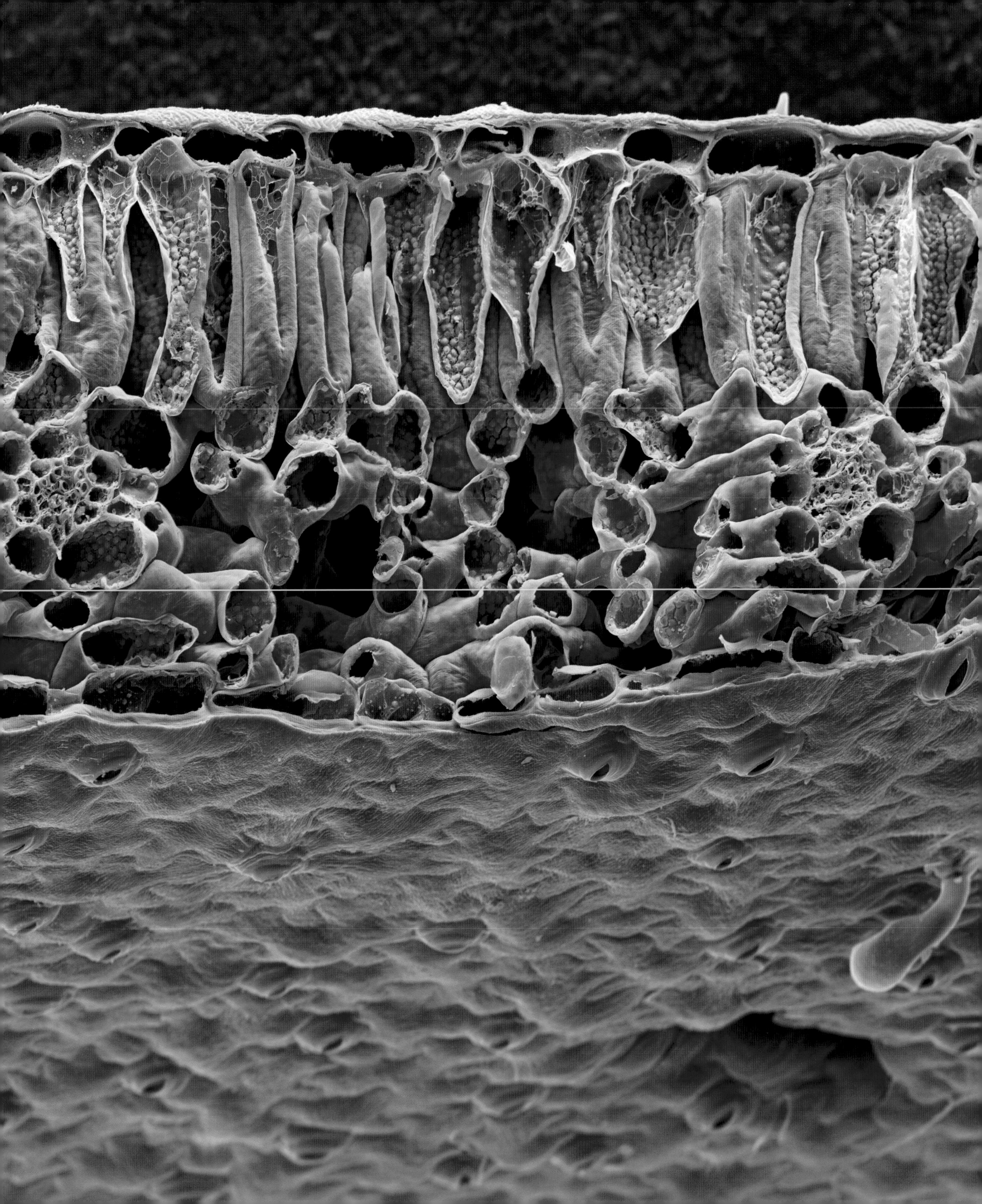

No two alike

the distinctive geometry of pollen grains

Biologists cannot agree about the purpose of sex. It may have evolved so that the genes of organisms are deliberately shuffled to keep one step ahead of chemically sophisticated viral and bacteriological parasites. Animals expend a great deal of energy on sex, often travelling vast distances to find mates. However, plants tend to be immobile (especially on land) and must mate at a distance. Pollen grains are the plant equivalent of male sperm in animals. Created in the stamens of flowers, they detach to drift in the wind, or attach themselves to insects – there are more than 20,000 species of pollen-carrying bee, for instance. Lured by sweet nectar in a flower, the insect becomes the instrument of pollination, depositing pollen grains on the female stigma of another flower. On arrival, the pollen grain divides into several cells, sends a tube through the receiving flower's stigma, and the next generation is seeded.

It is all very wonderful, but at least one in five people in the developed world associate the word 'pollen' with something entirely different – misery. Pollen grains can irritate the lining of the nose, causing the human body's immune system to kick into overdrive. Pollen allergies can cause considerable discomfort, and television weather forecasters routinely deliver pollen warnings during spring and summer.

Under the electron microscope, pollen grains reveal themselves to be unique for each species of flower, with an evolved design suited to whichever method of dispersal they favour. Grains carried by insects often have sticky shells which clump together as multi-packages whenever an opportunity for a ride presents itself. Wind-borne grains are extremely small, dry and light, like the finest talcum powder. Most grains have a hard outer shell (sporoderm) which is very difficult to digest. It is so durable, examples can be found in fossil deposits millions of years old. There are usually pores which allow the interior proteins and amino acids to interact with stigmas. They can also inadvertently interact with the human immune system – hence the runny noses and red eyes that so often accompany summer sunshine.

An assortment of pollen grains shows just a few of their immense variety of shapes. In this image, the grains appear to have been artfully lit, as if by a photographer. This is a typical effect of modern electron beam scanners, which can render the impression of directional lighting.

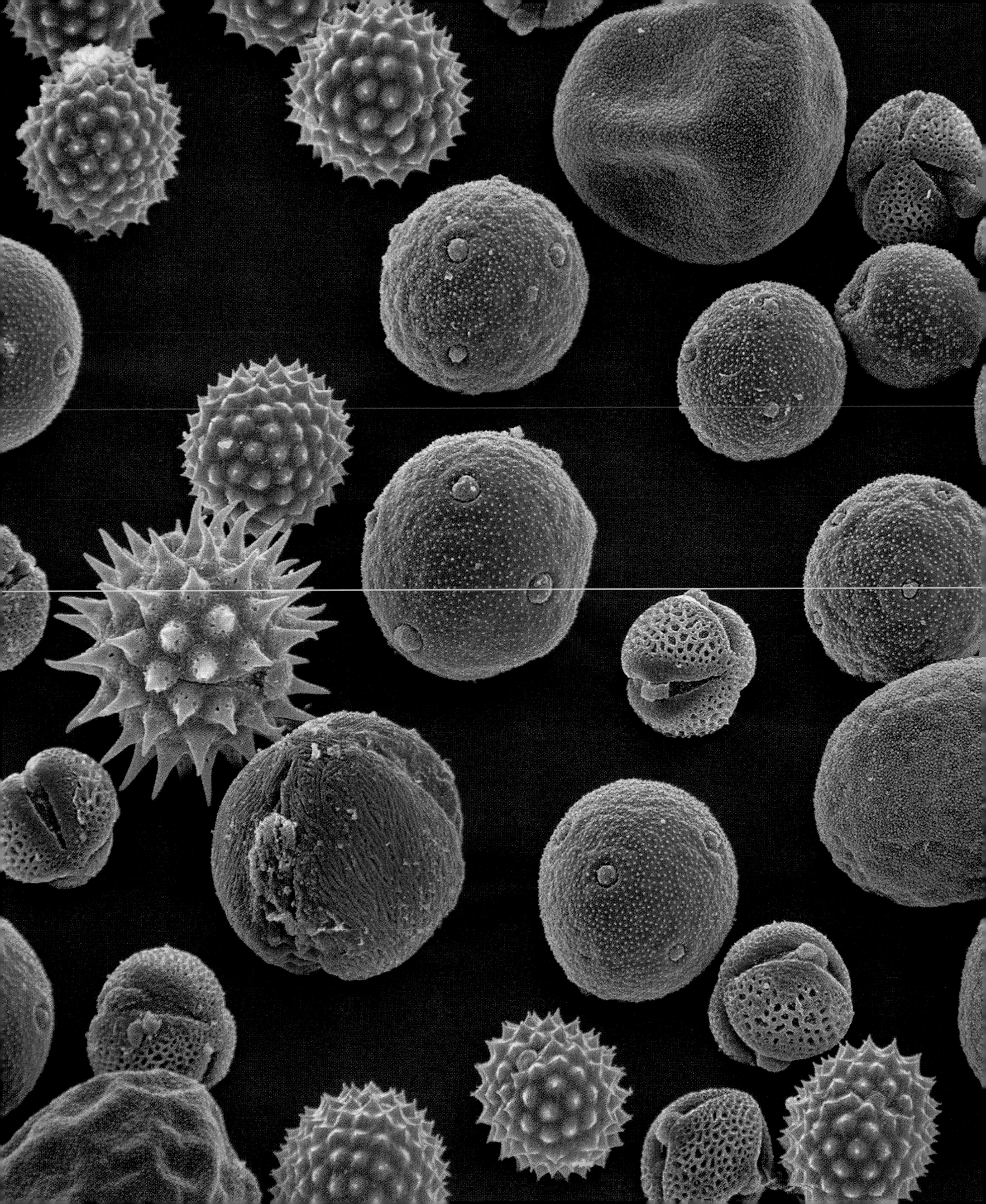

A monster menagerie

insects in extreme close-up

Something very like a three-dimensional view of an ant or a fly can be created in a scanning electron microscope. A beam of electrons, focused by magnets, sweeps across the surface of the ant. 'Secondary' electrons are knocked off the ant and are steered towards a phosphorescent television screen or detector to form an image. We have become familiar with a vast menagerie of insects and other small objects pictured in this way. But it is not easy to prepare biological subjects for an electron microscope, because the electrons usually have to travel through a near-vacuum (as in a television's cathode tube) before they impact with the specimen. Anything 'wet' will evaporate, and the shape of the specimen will become badly distorted. Samples need to be cryogenically dried before being placed inside the microscope, and then coated with a very fine vapour of gold or platinum to make them electrically responsive to electrons. The preparation of samples – even something as apparently 'crisp' as an ant or a mosquito – can be more challenging than the microscopy itself.

Insects may be small as individuals, but their impact on the world is truly staggering. There are at least 6 million different species. Given that ants and termites between them account for about 20 per cent of the mass of all animals on the planet, the total contribution of the insect class is probably about 30 per cent by weight. Ants and locusts can destroy farmers' livelihoods, but it is the malaria-carrying species of mosquito that have almost certainly had more impact on humanity than any other insect. Military campaigns throughout history have been decided in the bloodstreams of warriors rather than by the strategic skills of their generals. Even today, malaria claims one new victim every 15 seconds. We might wish to be rid of all mosquitoes, but at the same time as they plague us, their overall biomass (indeed, that of all insects) represents an essential food source for other animals. Electron microscope scans of insects are just part of a wide-ranging effort to understand them.

Drosophila fruit flies breed rapidly in captivity, allowing biologists working on genetics and evolution to study changes across many generations over a short time. The image below shows the mouthparts of a Rocky Mountain wood tick, which carries Colorado fever.

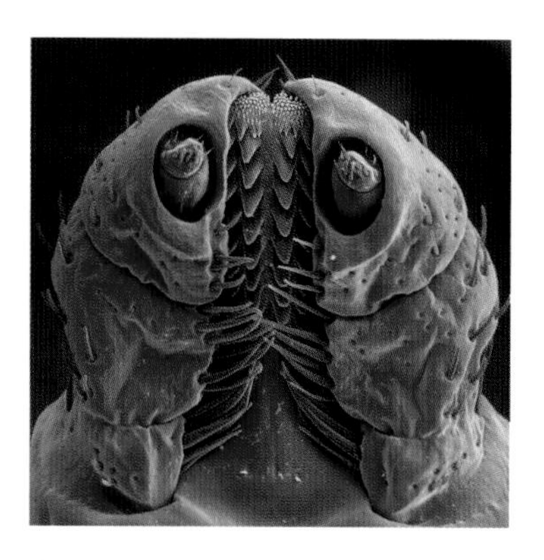

When science becomes art

false colours in electron microscopy

Today, much of the visual content of science comes direct from the laboratory instrumentation itself, and it is easy to assume that when electronic machinery delivers the images, the vague whims of human creativity are neatly excised and only pure, dispassionate science remains. Most of the images in this book appeared first on computer screens before being frozen for posterity in digital files at the touch of a keyboard. But arbitrary choices, settings on a dial, can sometimes adjust values of contrast and colour that are not necessarily set in stone by the underlying scientific data itself. Scientists may justify the fine-tuning of images on the basis that the presentation of information is being clarified, or specific features artificially highlighted, for ease of reference. Occasionally some confusion between art and science may result.

Some controversy has arisen over the enhancement of electron scans to make them more aesthetically appealing. The process works rather like a miniature radar or echo sounder, providing greyscale details of surface texture and shape, but not of colour. Yet many electron scans are published with colours defining different features: the eyes and legs of insects, for instance, or the difference between the bugs and the surfaces on which they rest. So sophisticated has this process become (especially with modern software tools), some electron microscopists now think of themselves as artists just as much as scientists, and market or display their work accordingly. This does not always sit so well with scientists keen to preserve a strict demarcation between reliable data and subjective interpretation.

Can science be art? In the case of electron microscopists adding colours to their otherwise greyscale images of bugs and pollen grains, the answer is approximately 'yes'. Tina Carvalho of the University of Hawaii's electron microscope facility is far from alone when she admits, 'I use Adobe Photoshop to paint my insect images with a mouse or digital pen and pad, sometimes working in great detail, on the level of individual hairs, even. The colours are a product of my imagination, but they do help show the subjects more clearly.' Even something so straightforward as the angle from which a specimen is scanned may be determined by aesthetic choices as well as the needs of science.

An extreme close-up of the ball and socket joint at the base of an ant's feeler has been coloured for visual effect. The malaria-carrying mosquito (below) has also been artificially coloured, especially to highlight its compound eyes.

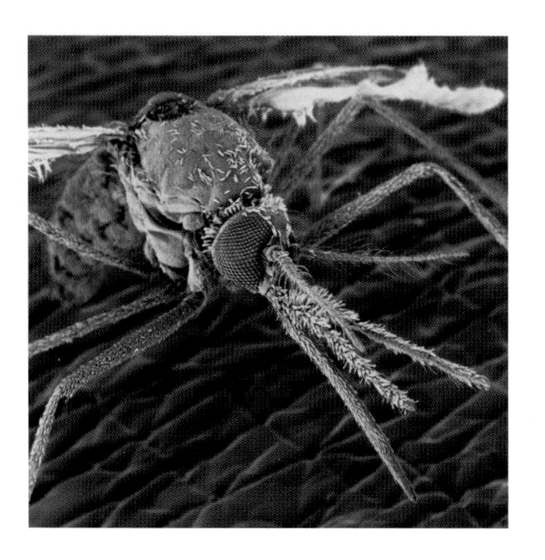

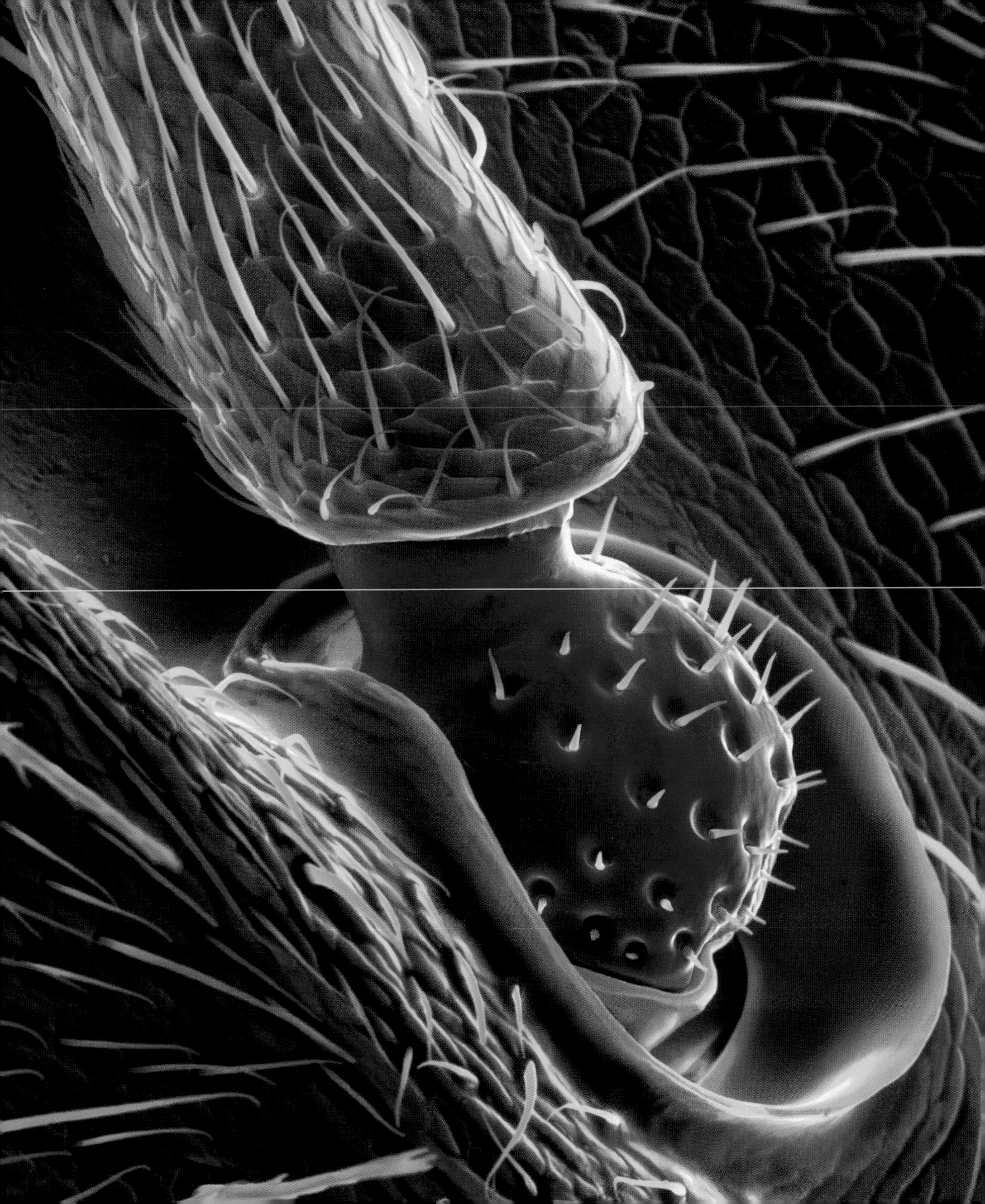

A disconcerting ecology
the secret life of dust

We seldom stop to think about what dust really is. To the unaided eye, it just seems a simple grey powder, an everyday nuisance, but closer inspection under the microscope reveals some unsavoury details. Most household dust turns out to be pieces of us: our hair, our skin and our clothes. We constantly shed dead skin cells, which drift in the air for a while and eventually settle on floors, tabletops and any other flat surfaces in the house. Fibres from even our best-quality clothes spontaneously fall out. The outside world contributes too. Dry fungal spores, pollen, mineral grains from soil, and above all the sooty carbon deposits from vehicle exhausts, all add to the mire. Finally, our houses come apart, shedding their own 'dead' skin. Floors wear away slightly every time we walk on them, leaving powdery traces. Paint comes off the walls, and antique tables drop shavings from the woodworm's secret endeavours.

Amidst all this delicious debris, 'dust mites' thrive. They are not insects but are more closely related to spiders and ticks. They live in bedding, couches, carpets, stuffed toys (our favourite teddy bears) and old clothes. They feed on the dead skin that falls off the bodies of humans and household pets, and on any other organic material around where they live – sometimes each other. A clump of dust can be a happy hunting ground for mites that feed on other mites. Their droppings, and the skins they shed every time they put on a spurt of growth, drift in the air whenever we vacuum a carpet or dust the tabletop in our efforts to keep a clean house. It is now believed that mites and their debris are a primary contributor to childhood asthma and other allergies.

Tiny bedbugs inhabit our mattresses and blankets, and when we shake our sheets out in the morning, we scatter them and their droppings into the overall collection of dust. Bedbugs feed on human blood, but fortunately, there is no evidence that they can transfer diseases. Because of their small size and inconspicuous nature, bedbugs can be transported from one house to another in furniture, clothing and laundry. Once in a house, they can hide almost anywhere: under the floorboards, even behind wallpaper. One might think they could be starved out of a house by simply leaving them without food for a few days, but no; they can survive more than a year without feeding.

Dust mites exist in a universe of such tiny scale, we are seldom consciously aware of their existence; yet they surround us in vast numbers. Household dust (below) contains hairs, woollen fibres, plant debris, and scales and excreta from mites and insects.

The traffic inside us

blood cells in unprecedented detail

To the naked eye, blood is a uniformly red fluid. In fact nearly 90 per cent of it is clear water. Under magnification we see the individual, highly complex swarm of worker cells that moves through our body, carrying energy and fighting infections.

The flat, disc-shaped red cells, also known as erythrocytes, are the most abundant blood cells, giving it that characteristic red colour. Their main job is to transport oxygen from the lungs to the rest of the body, and then carry waste carbon dioxide back to the lungs for exhalation. There are about 5 million red cells in a cubic millimetre of blood.

White blood cells (of which there are many complex varieties in a healthy body) are the most important components of the immune system. Properly known as leukocytes, they are unlike other cells in the body. They act like independent living organisms, able to move about and capture prey, engulfing bacteria and other invasive agents whenever cued by their biochemical alienness. Our bloodstream is a vicious war zone, where white cells seek and destroy dangerous enemies. Hunter and hunted constantly adapt, and the battle is never decisively won. Sometimes, however, it can be decisively lost.

Neutrophils are by far the most common form of white blood cell. Our bone marrow produces billions of them every day and releases them into the bloodstream, but their life span is short: generally less than a day. They move through capillary walls into tissue, where they are attracted to foreign material and bacteria. On encountering an enemy, they engulf it, releasing enzymes, hydrogen peroxide and other potent chemicals. In a site of serious infection, pus will form. Pus consists mainly of exhausted neutrophils and the wreckage of defeated infections. No wonder it is so often associated in our culture with the stench of bodily decay, but it is part of a valuable process.

Because white blood cells are so important to the immune system, they are used as a measure of health. When you hear that someone has a 'strong' or a 'weakened' immune response, one way this is determined is by counting all the different types of white blood cell in a blood sample.

Blood is not merely a fluid. It is an amazing traffic system, a carefully balanced energy economy, and a river of relentless battle.

To the unaided eye, blood appears a deep crimson red. Under an electron microscope, however, the 'redness' is revealed as belonging to just a few types of cell. The fluid plasma that makes up most of the volume of blood is entirely transparent.

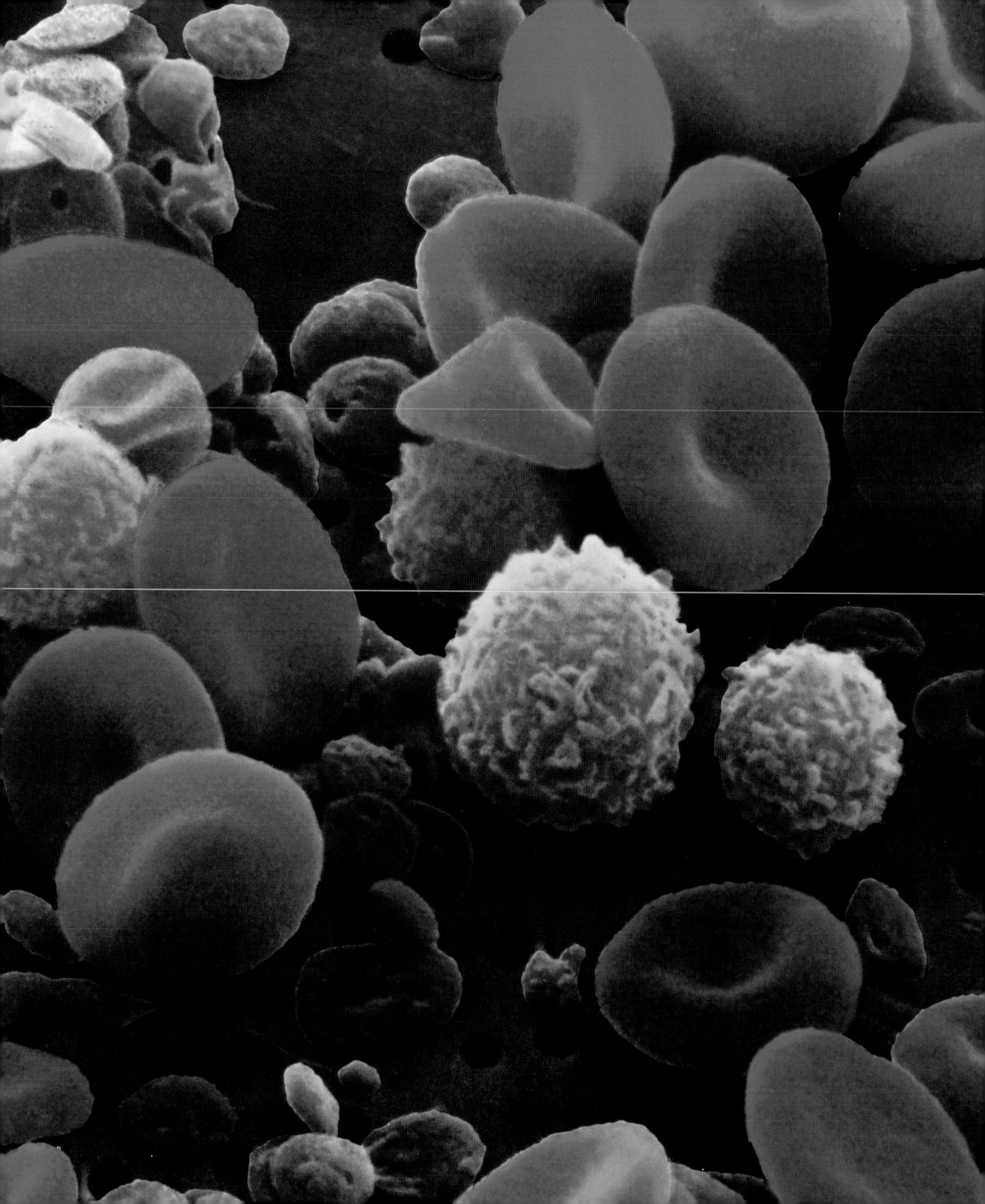

Capturing the essence
living samples under the scanner

Conventional electron microscopes can only study dried samples in a near-vacuum. Biological tissues need to be freeze-dried to remove almost all their water content, so that they do not burst apart once the vacuum is applied and the water boils away. Then a fine vapour of gold or platinum atoms is introduced into the sample chamber, and the vapour condenses on the surface of the samples, until they are entirely coated in a thin veneer of metal. This ensures a clean, sharply defined reflection of incoming electron beams. The resulting level of detail can be quite staggering, but the images retrieved lack something of the basic truth of biology: aliveness.

The latest environmental scanning electron microscope (ESEM) allows samples to be protected from decompression by surrounding them with water vapour or ordinary air. They can even be kept alive during a scan. The primary electron beam hits the specimen, which emits secondary electrons. As they fly through the gaseous environment, they collide with atoms in the gas, knocking off more electrons. This increase in the overall number of electrons in the return pulses effectively amplifies the signals. For the first time, we can see whole cells and living entities at work, and not merely their dessicated metallic shadows.

The stomata of leaves can be observed opening and closing; insects can be seen in motion; wet nerve and muscle tissue can be examined; and the absorption of drugs through cell membranes can be studied as it happens. Biology is not a catalogue of inert things – it is a complex set of *processes*, and ESEM machines enable them to be observed. The one drawback with ESEM is that the degrees of magnification are not so great as with conventional vacuum electron microscopes. In the near future, no doubt ESEM resolutions will improve. For now, at least we can begin to witness the dynamics of life in action, rather than merely the dessicated remains of life brought to a halt.

Two cervical cancer cells form during division from one cell. They are still joined at the centre. Typically, the cells have an uneven surface, and they divide rapidly in a chaotic manner. The smaller image (below) shows healthy cells from a heart, bundled into muscle fibres.

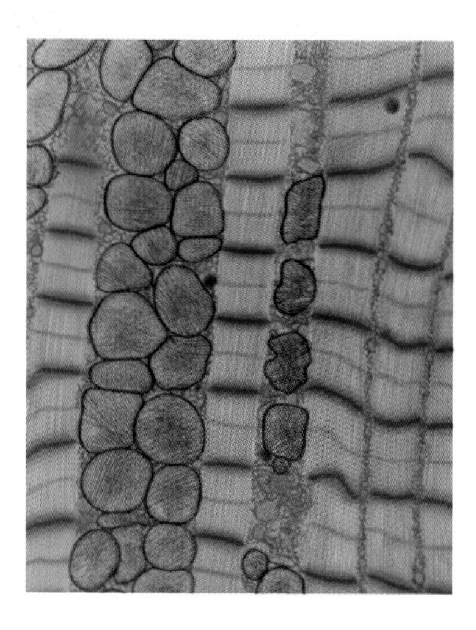

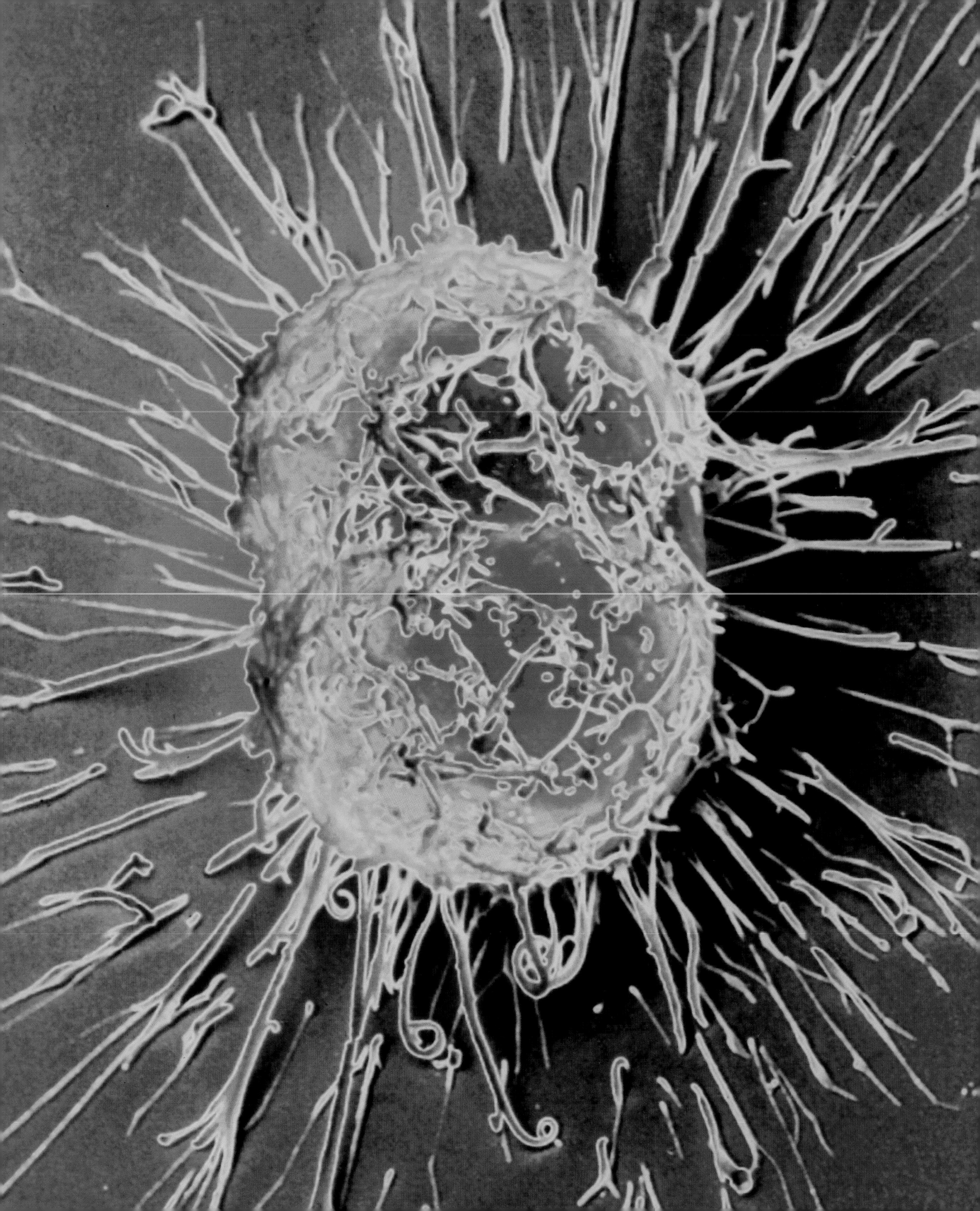

Electron illusions

microscopic life from Mars?

Around 15 million years ago, a fist-sized chunk of rock was blasted from the surface of Mars during a powerful meteorite collision. After drifting through space, it crossed the Earth's orbit and fell on Antarctica, where it lay hidden under the ice for 13,000 years – until its discovery by a US survey team in 1984. In the summer of 1996, a team of NASA scientists, led by David McKay, announced the discovery of what they believed were fossilised traces of life in the meteorite, known as ALH84001. Their most controversial piece of evidence was not included in their initial paper for *Science*; it seemed altogether too fantastic. At the extreme magnifications delivered by their scanning electron microscope, McKay's team picked out physical structures resembling fossilised bacteria. They were teardrop-shaped, and some of the examples even appeared to be segmented. The trouble was, these 'fossils' were smaller than any bacteria yet found on Earth. At least, so it seemed to critics at the time. A few weeks later, researchers in Washington State announced they had found terrestrial bacteria just twice the size of the supposed Martian fossils.

There comes a point where ambiguities at extremely small scales can affect the processes of electron microscopy. Other teams looked not at bacteria or biology, but the preparations of ALH84001 samples prior to the microscopy. In order to generate a distinct image, McKay and his team had followed a perfectly standard procedure of injecting gold vapour into their sample chamber, so that a fine layer of metal would coat the samples, thus creating a more defined reflection of incoming electrons. Sceptics ran tests to see if any interesting accidental shapes could be created from unevennesses in the gold coating process – and sure enough, vaguely bug-like gobbets emerged in some of their experiments.

At these tiny scales, every nuance of chemistry, surface texture and instrumentation counts. McKay and his team replied, somewhat tersely, that of course they had run control experiments to check their gold deposition process; and there were so many other chemical and mineral clues associated with their images, it was unlikely that they had come up merely with an 'instrumental artefact'. Whether there were bacteria in the Martian meteorite, or the evidence was a result of the scientists' technique, is as yet undecided. The jury remains divided over one of the most fascinating questions of our time.

The segmented, worm-like structure shown here, one hundredth of the width of a human hair, may be the microfossil of a primitive bacterium that existed on Mars 3.6 billion years ago. It may also be a random mineral formation, or a product of the techniques used to make the scan.

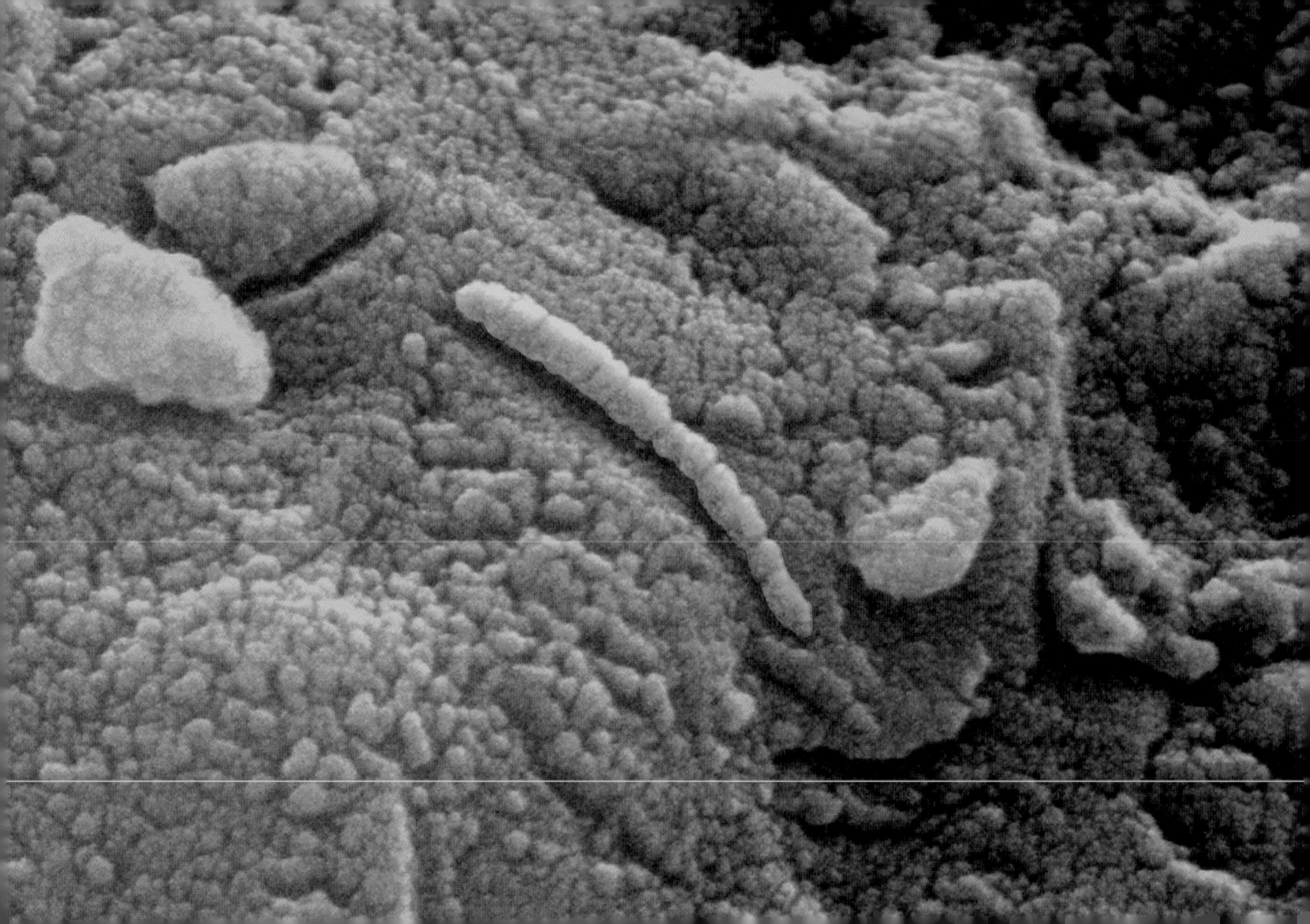

Anatomy without knives

For many centuries, we could learn about our bodies only by slicing into other people's. Even in recent decades, diagnosis of serious illnesses often required surgery just to identify specific problems. Today, we have incredible scanners that fulfil the wildest predictions of science fantasy. We can locate tumours, pinpoint flaws in muscle and bone, and monitor the flow of blood through veins and arteries, without the need to put patients under the knife. We can even map electrical and chemical impulses surging through the brain. The human body is becoming transparent as never before.

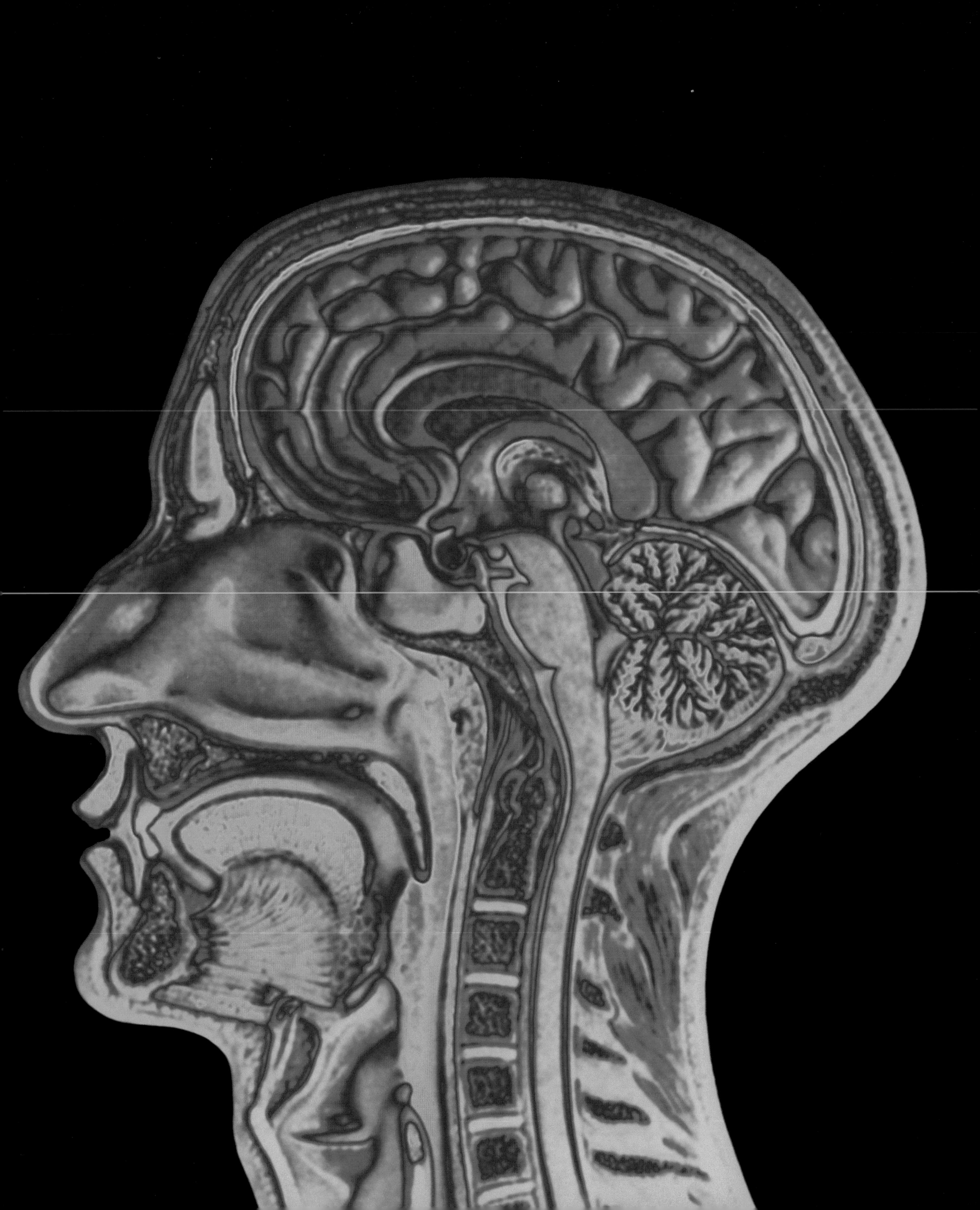

The X-ray
a century of medical successes

As with many great inventions, X-ray technology was discovered by accident. In November 1895, German physicist Wilhelm Röntgen was experimenting with 'cathode rays' (electron beams) emitted from a hot metal filament inside a glass vacuum tube. He noticed that a fluorescent screen at the other end of his workbench started to glow. This response in itself was not so surprising – fluorescent material painted onto a sheet of cardboard normally glowed in reaction to the rays emitted by vacuum tubes – but the tube was swathed in black cardboard on this occasion, and Röntgen assumed this would have prevented any rays from escaping. When he placed slightly thicker card between the tube and the screen, the screen still glowed. Finally, he put his hand in front of the tube, and saw the silhouette of his bones projected onto the screen. Röntgen had no way of knowing that the electrons in his vacuum tube, generated by a particularly hot filament, were so energetic they were generating an altogether different kind of radiation when they knocked into atoms in the glass walls of the tube.

A week after discovering X-rays themselves, Röntgen also discovered their most beneficial application. He captured an image of his wife's hand on a photographic plate. Her bones (and her wedding ring) showed up as solid shadows. She was horrified by the death-like image, but the medical world quickly embraced this new discovery, which seemed little short of magic. Here was a tool that could look inside a living patient and reveal broken bones, or seek out the exact location of a shard of shrapnel prior to surgery. But what *were* these rays? An invisible kind of light? Something akin to the cathode tube energies? At first, no one was sure, and this is why they were called 'X' rays.

Now we understand X-rays to be energetic short-wave radiation, which can be highly penetrative and disruptive to organic molecules. Exposure to the rays increases the statistical chance of DNA or other cellular damage, which in turn may lead to cancers. But this is a statistical rather than an absolute risk. For over a century, the medical profession has weighed these risks against the huge benefits of being able to look inside the human body without cutting into flesh. Short-term exposure to X-rays is widely regarded as acceptable for patients.

PRECEDING PAGES
A magnetic resonance imaging (MRI) scan of a healthy male human head creates a perfect cross-sectional view, showing most of its major structures: the brain and brain stem, the neck vertebrae, the tongue, the sinuses and the vocal chords.

FACING AND BELOW
A delicately colour-enhanced X-ray of a knee joint reveals three bones: the femur (thigh bone) the top of the tibia (shin bone) and the patella (knee cap). The small image is Röntgen's original 1895 X-ray of his wife's hand. Notice how her metal ring casts a strong shadow.

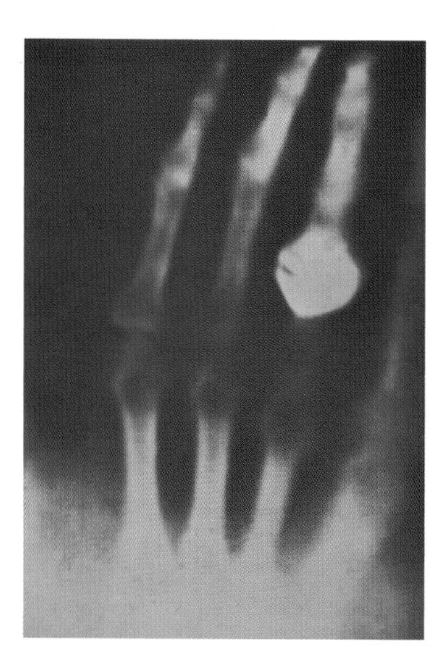

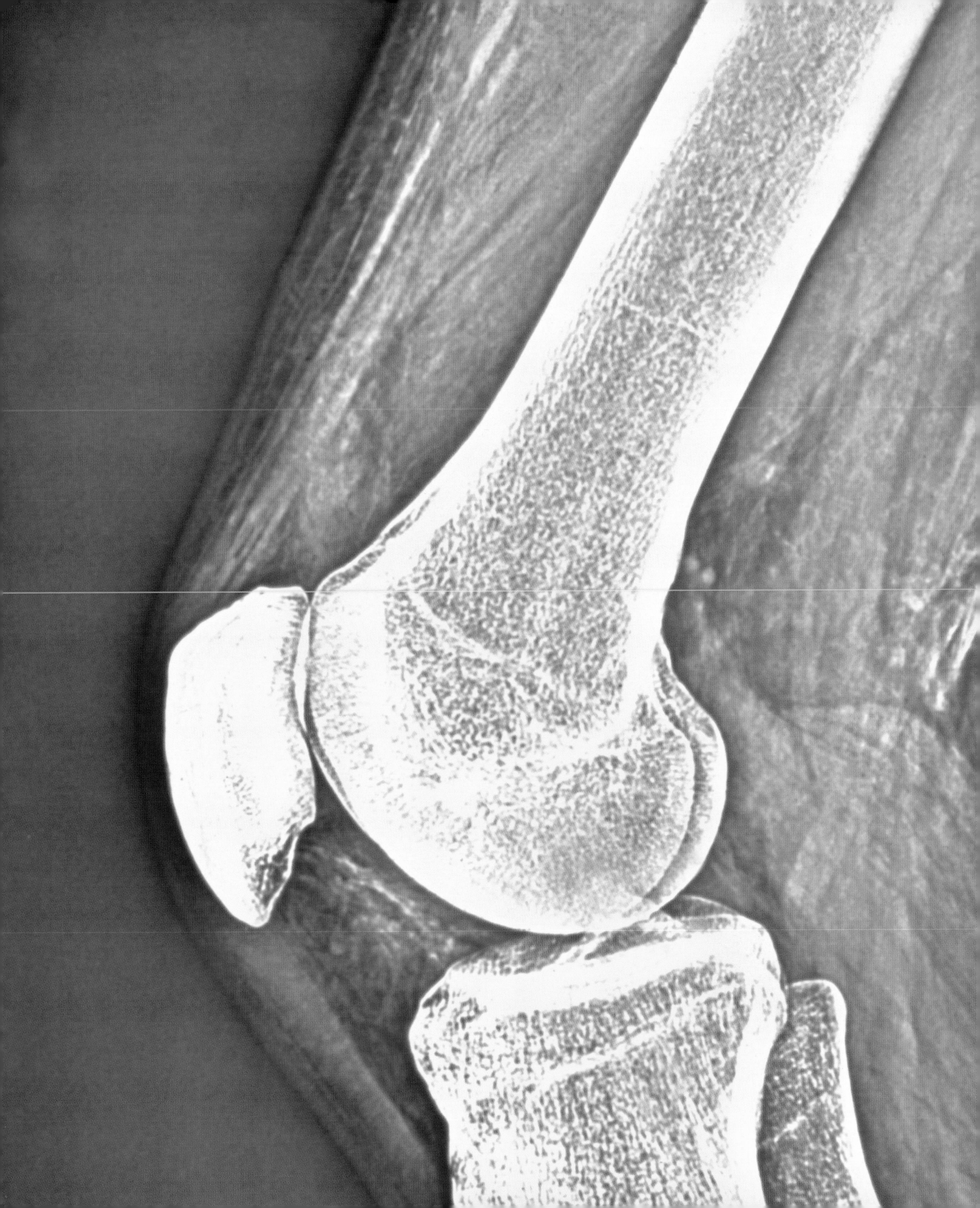

Painting over the cracks
works of art dissected by radiation

When artists present their work to the world, they may not mean us to see all the indecisions and revisions that played a part in their achievements. Film makers cut unwanted sequences and leave them on the editing room floor. Novelists discard rough drafts, and painters assume that the top surface of their canvases will stand for all time as the definitive displays of their artistry. It is not always that way these days. Half-baked manuscripts, once hidden in desk drawers, find the light of day when the inheritors of a famous novelist's estate see a profitable opportunity. Movies marketed on CD are replete, now, with scenes that were discarded from the main cut of a film by a director, yet which the studios want to exploit; and the intentions of famous old masters at their oil painting easels are revealed to us, whether or not that is what they might have liked. X-ray, infrared and ultraviolet scans strip away the surface appearance of the paint to show the layers of workings beneath, from rough charcoal sketches outlining an idea on blank canvas, through to entirely different versions of a painting applied on top of older, unwanted compositions.

Ultraviolet light is (almost always) invisible to the naked eye, but it can cause many substances to fluoresce by exciting electrons so that they emit photons of visible light. Characteristic fluorescences reveal the presence of particular chemicals in paint pigments, while the *absence* of expected responses from a supposedly old painting may reveal the contributions of restorers (or outright forgers) working long after the original work was supposed to have been completed, using paints that did not exist in the attributed artist's lifetime. Similarly, the components of a painting examined under X-rays will show varying absorption characteristics. Pigments containing lead or mercury, for instance, will block more rays than those containing chromium or cobalt. Careful tuning of the emitted X-ray frequencies can strip away different layers of a painting, and few of the secrets of its construction remain secret for very long. Sometimes we see the bravery and talent of a great artist as he assembles his masterwork. On other occasions, we see ghastly mistakes overpainted by someone who must have been pretty confident that no one would ever discover how badly his first version came out.

A portrait of a young woman (below) by the Spanish artist Goya (1746-1828) is more than it might appear in daylight. Under X-ray photography, it becomes clear that it was painted over a completely different picture. The scan was made at the National Gallery in London.

Slicing the subject

computerised axial tomography

Today we benefit from many scanning devices based on the general principle of an X-ray machine, but harnessed to powerful imaging computers, and emitting and detecting a variety of radiations – some of which are so exotic they almost defy common-sense imagination. We can now discern much about a person's health without the need for exploratory surgery. The dream of a *Star Trek*-style universal health scanner may not be that far away.

In 1972, Godfrey Hounsfield, at the British company EMI, unveiled a new technology: an X-ray machine allied to a computer. This device, now universally known as a CAT scanner (for computerised axial tomography), has become one of the most valuable diagnostic tools since the discovery of X-rays themselves. Instead of shooting and processing X-ray photographs one at a time, a CAT's X-ray source is rotated around the body, and then ratcheted along and rotated again, to deliver a series of slice-like scans. These are then compiled and assembled as a coherent picture on a screen by the computer. CAT does not deliver irreversibly fixed photographic images; rather, it provides a fluid set of data that can be translated and viewed in many different ways.

A conventional X-ray photograph is a two-dimensional shadow derived from a three-dimensional subject. If one bone lies in front of another relative to the X-ray machine's point of view, that second bone will be hidden from view and will not show up on the plate. CAT scanning allows a 360-degree tour to be taken of the target area, rotating the view so that no one object can hide behind another for very long.

Modern CAT scanners often generate startlingly vivid colour images. The colours have nothing to do with the real appearance of body organs; they are a means of coding different tissue types and densities, defined by their relative transparency or opacity in front the scanners' X-rays. Unfortunately, those rays are still as dangerous now as they have always been. X-ray technology of any kind requires that patients accept a very slight statistical risk of developing cancers later in life. So far, the immediate benefits of CAT scanning have almost invariably outweighed the longer-term risks.

A transverse section of a human torso, captured by a CAT scanner working with X-rays, reveals mesothelioma cancer (highlighted in red) constricting the right-hand lung. The disease was almost certainly triggered by exposure to asbestos. The other lung is healthy.

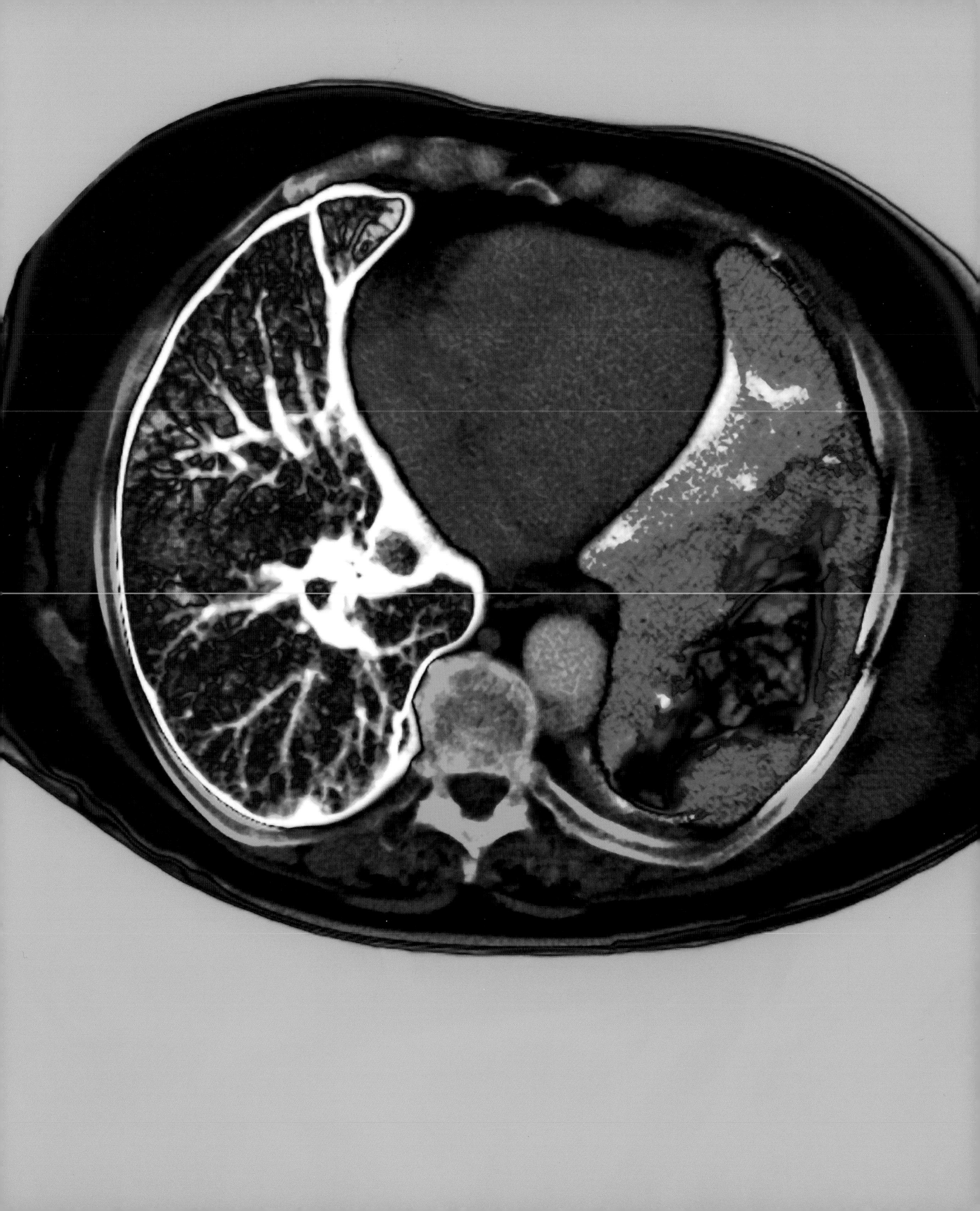

Archaeology without tears

scans of delicate artefacts

If there is one thing that irritates a typical investigator of fragile fossils or ancient artefacts, it is the botched interventions of previous investigators. The very process of archaeology runs the risk of compromising the physical properties of a sample in some way. Wrecked wooden boats, preserved for centuries in estuary muds, can fall apart in days when exposed to the open air. Broken shards of pottery might be glued together by one generation of restorers, only for the next generation to discover that the glue is stronger than the shards, and cannot safely be removed. The early twentieth-century discoverers of Tutankhamun tore his mummified remains apart, seeking to establish the cause of his death, and in so doing, they destroyed far more clues than they revealed. Time and again, modern archaeologists have cause to regret the work of their predecessors. Aware that future generations might find our tinkering with valuable samples just as clumsy, the emphasis is now on trying to do as little as possible to any delicate ancient object that we encounter. If we can see it – look inside it even – without actually touching it, so much the better.

CAT and MRI scanners were originally created for use in hospitals, but archaeologists find them valuable for exactly the same reason as doctors: they can probe 'patients' without the need for dangerously invasive cutting. A fossilised dinosaur egg, for instance, can only be inspected visually by slicing into its hardened, mineralised crust. Scanning technology allows the shadowed foetal form inside to be modelled as a three-dimensional graphic, leaving the real egg completely unharmed.

Fossil experts are lucky if they find unbroken dinosaur eggs, or the skeletons and skulls of adult creatures intact. It is not only that a fossil may be tens of millions of years old, and fragile as a cobweb; in all likelihood, the animal in question died in the first place because it was torn apart by a predator; and even if it expired peacefully of old age, its carcass would probably have been dismantled by opportunistic scavengers. Modern, three-dimensional scanners are valuable to researchers when faced with the crunched remains of prehistoric meals. The outlines and contours of missing sections of bone can be inferred by computer. Synthetic components are then moulded to fit the gaps, and something very close in form to the original skeleton is recreated. There is nothing new about replacing lost bone fragments with artificial fillers, but computer analysis speeds up the process and cuts out unnecessary handling of fragile samples.

A computer image of the skull of a triceratops, created from scans of actual fossil bone fragments, and completed by extrapolating the shapes of missing bone pieces. These techniques reduce the need for handling real objects, which may be fragile.

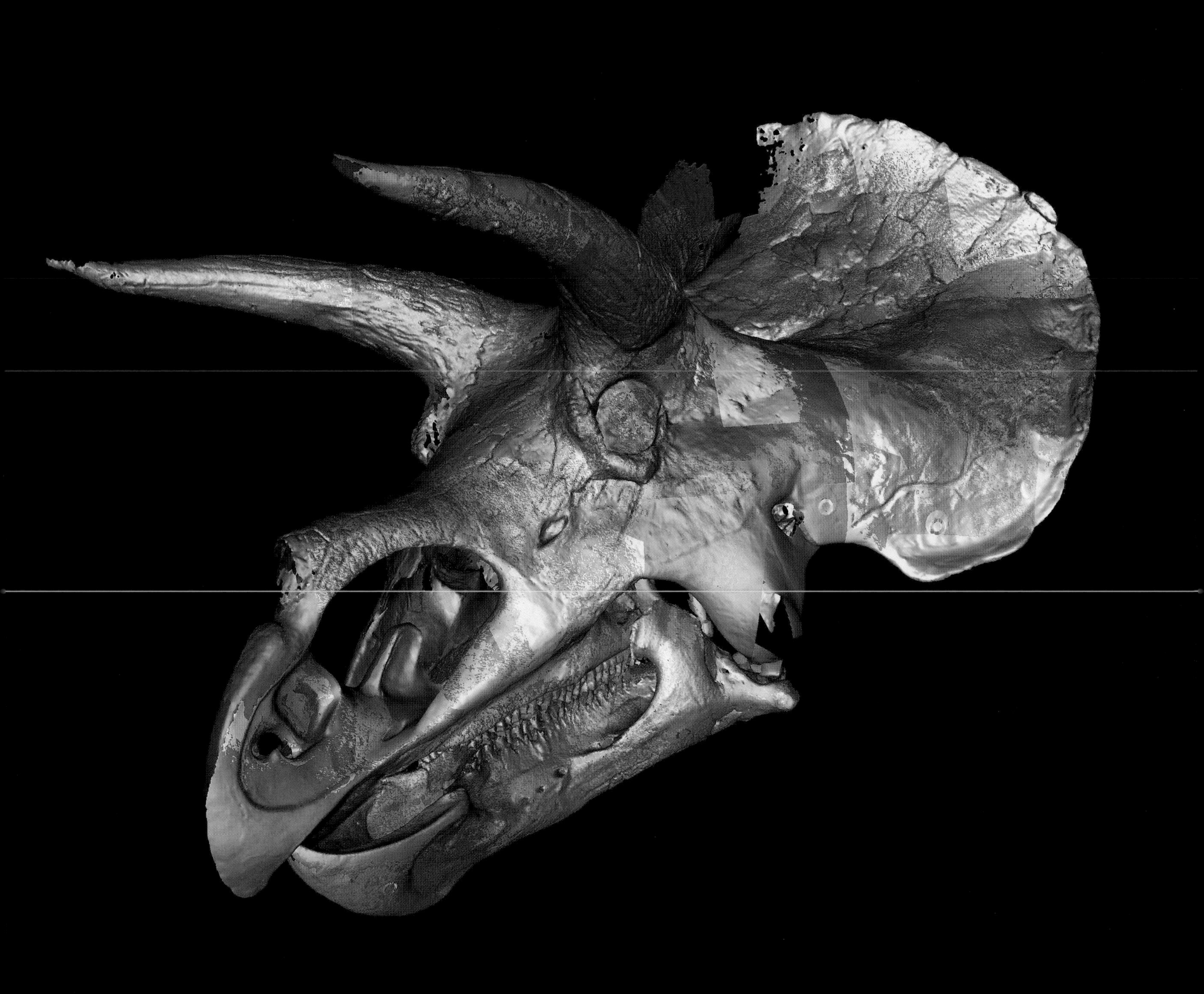

The Tutankhamun case

an ancient murder probed with new tools

In 1925, Howard Carter was allowed to open the sarcophagus of Tutankhamun, the ancient Egyptian boy-king whose tomb he had famously unearthed three years earlier. Dr Douglas Derry performed a crude post-mortem, but the conditions were hardly ideal: Tutankhamun's mummified body was sealed in a gelatinous mass of resins, which had all but solidified in the 3,300 years since his burial; and the X-ray machine specially shipped to Egypt for the occasion broke down. In 1969, a British team headed by R. G. Harrison obtained permission to X-ray the mummy. They saw what Derry had missed: that the sternum and most of the ribcage were missing. They must have been removed during the embalming process, but why? X-rays of the skull also showed two small chips of bone suspended in the resins that had been poured into the brain cavity by the embalmers. Were they significant?

In 2002, Todd Grey, Chief Medical Examiner for Salt Lake City, took an interest in the 1969 X-rays. He had seen many similar skull injuries in his career. As a falling person hits the ground, the brain is thrust forward and hits the front of the skull. These kinds of shock can loosen delicate internal bone fragments, especially from areas around the nose and eye sockets. Grey thought it was possible that the subsequent ritual removal of Tutankhamun's brain matter, and replacement of it with embalming fluids, could have moved the loose bone fragments deeper into the brain cavity. He concluded that Tutankhamun might have been murdered by a blow to the head.

What about the missing ribcage? It was hard to imagine a murder method that involved crushing a ribcage so badly that embalmers had to dispose of it in order to tidy up the body. A hunting accident seemed more likely: a fall from a chariot, perhaps, after which the hapless king landed heavily on his chest. The bone fragments in the skull were probably no more than incidental products of the embalming process and perhaps, even, of Dr Derry's controversially destructive post-mortem examination in 1925.

The Tutankhamun case is one of ambiguities and speculation, because so much evidence is missing from the story. X-ray scanners can tell us in amazing detail the state of an artefact at the time of a scan. How it got to be that way, though, is not always so easy to discern.

R. G. Harrison's 1969 X-ray plate of Tutankhamun's mummified head. He concluded that the young pharaoh 'could have died from a brain haemorrhage caused by a blow to the skull from a blunt instrument'. In 2002 this claim was refuted; the debate continues.

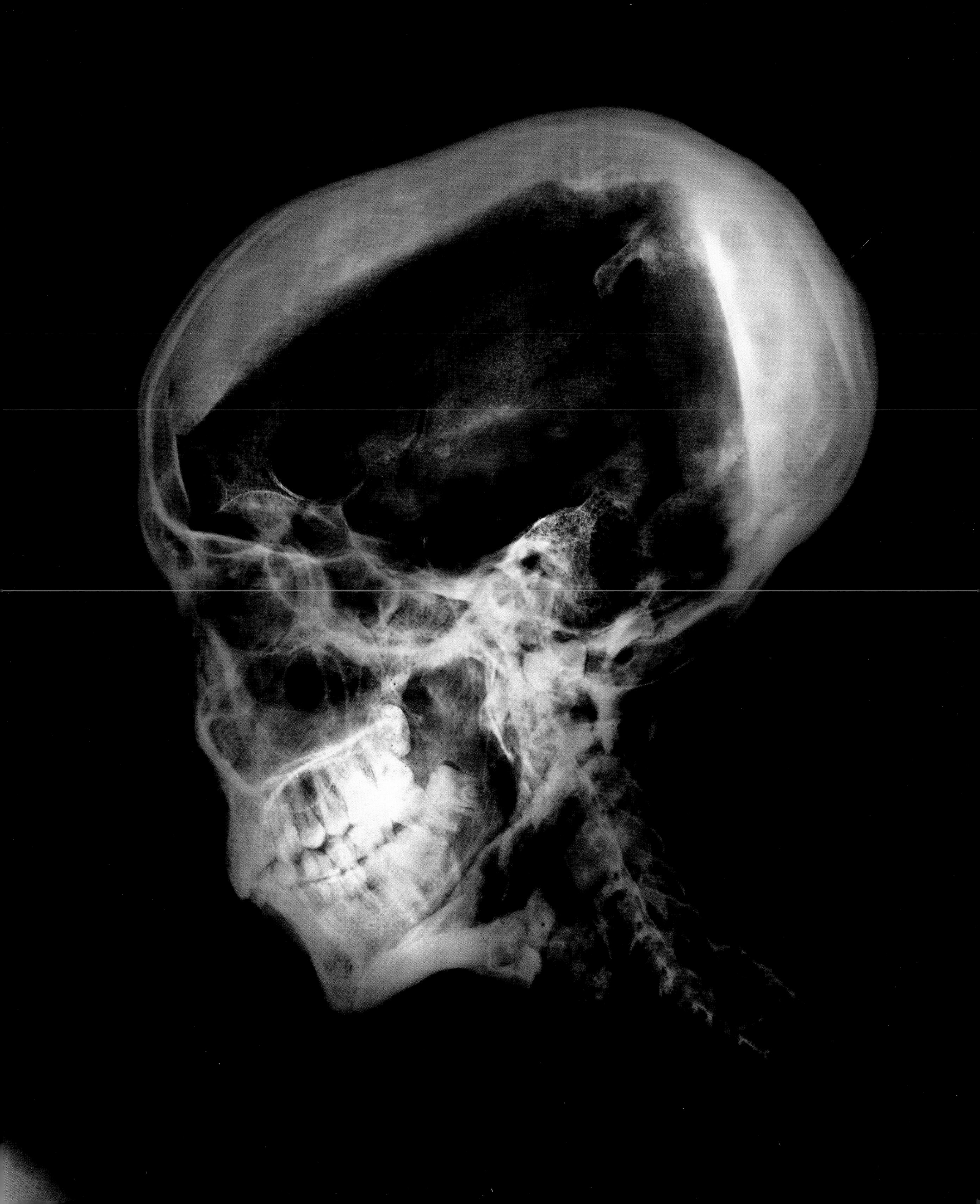

Seeing like a dolphin
ultrasound images of the human body

In 1958, Professor Ian Donald at Glasgow University saved a woman's life with sound: he proved that her cancerous tumour was nothing more than a harmless cyst, thus obviating the need for the drastic treatments his colleagues had in mind. Donald adapted a machine whose principal function was to seek out flaws in metal welds, believing that it might just as well identify flaws in people. The medical profession as a whole soon agreed with him.

Ultrasound scanners transmit high-frequency sound pulses into the body. When the pulses hit a boundary between tissues (for instance, between fluid and soft tissue, or soft tissue and bone), some of the waves are reflected back to the probe. The scanner then calculates the distance from the probe to the tissue boundaries, using the speed of sound in tissue, and the time of each echo's return – usually in millionths of a second.

Sound waves are slow and clumsy in comparison to high-frequency electromagnetic radiation. Even the highest-pitched ultrasound devices cannot distinguish very small structures in the body – but they do have one overwhelming advantage. Conventional X-ray scans always pose a modest risk to patients, and unborn babies in particular are susceptible to genetic damage. Sound waves, on the other hand, are harmless, and so ultrasound machines have become a familiar component of prenatal care. A small hand-held probe, placed in contact with the abdomen, transmits and detects high-frequency sound pulses in a wide arc, and the sound energies are absorbed or reflected by different tissue types.

Recent research indicates that dolphins can also 'see' some of the internal boundary layers of objects they probe with their high-frequency sounds. Distinctively shaped toys hidden inside other toys with different shapes make an easy test, especially when dolphins are rewarded with food treats for finding them. There are also indications that dolphins can tell when other dolphins are pregnant, or even ill. Ultrasound scanners are just one of many examples of human ingenuity that are the merest shadows of pre-existing natural phenomena.

Millions of expectant mothers have experienced ultrasound scans, in which an arc of fuzzy light on a monitor screen provides an approximate image of the unborn child. Modern techniques allow more distinct images to be made, and even 'portraits' to be created.

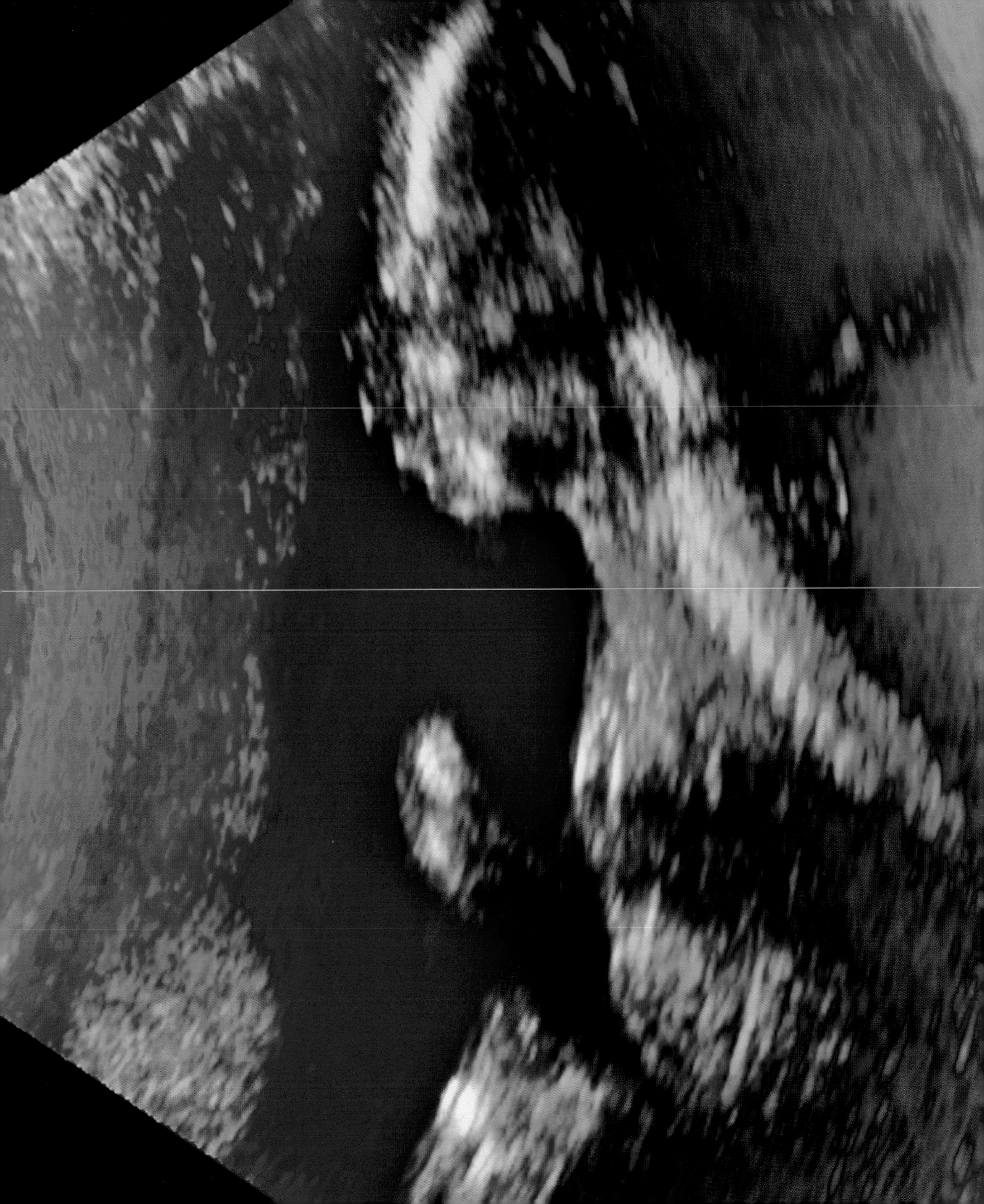

Reading the human code

do genes control our behaviour?

A strand of DNA is arranged like a spiral staircase with two bannisters twisting around each other. The 'rungs' of that staircase are built from just four simple compounds: cytosine (C), guanine (G), adenine (A) and thymine (T). Normally, C will link to G, and A to T. Immensely long sequences of these 'base pairs' store the biological characteristics of an organism, much as a digital computer stores a program. Over the last two decades, genetic scientists on both sides of the Atlantic have analysed human DNA, in one of the most intensive scientific projects ever attempted. Competition between America and Britain was especially intense at first, until all the scientists, whether publicly funded or commercially driven, at last decided to swap notes and combine their haphazard collections of fragmentary data. The 'Human Genome Project' finally delivered a complete list of all the 3 billion base pairs in human DNA. The work had taken 15 years.

Can we now 'see' how a person works by uncoding their DNA? The genome research has been impressive, and will yield tremendous benefits in terms of our overall understanding of the broader workings of DNA – but each of the 6 billion people on the planet has slightly different DNA, because of the recombinations that result from the sexual unions of their parents. If a criminal suspect, for instance, is identified by 'DNA profiling', the chances of their being mistaken for someone else are one in a billion. Legal arguments usually centre on failed record-keeping in police departments, or ambiguities about the sample collection at crimes scenes, rather than the actual analyses of DNA. Few people seriously doubt that DNA defines our hair colour, physical build and blood type. It also disposes us towards (though by no means utterly predetermines) certain illnesses, and protects us from others. But what about our characters, our likes and dislikes, our leanings towards honesty or criminality? Does DNA determine our personalities? Will we one day be able to 'read' the code for humanity? This is an extremely contentious issue. Evolutionary biologists argue that many traits in humans, such as ambition, sex drive, competitiveness or tenderness towards loved ones and close friends, are so prevalent they must have logical genetic explanations. Others argue that strands of chemicals cannot fully explain who we are, nor determine the conscious choices we make.

A computer display of a human DNA sequence represented as a series of coloured bands, with the colours standing in for just four simple compounds that carry DNA's information content. Can all our human qualities be described as a product of that code?

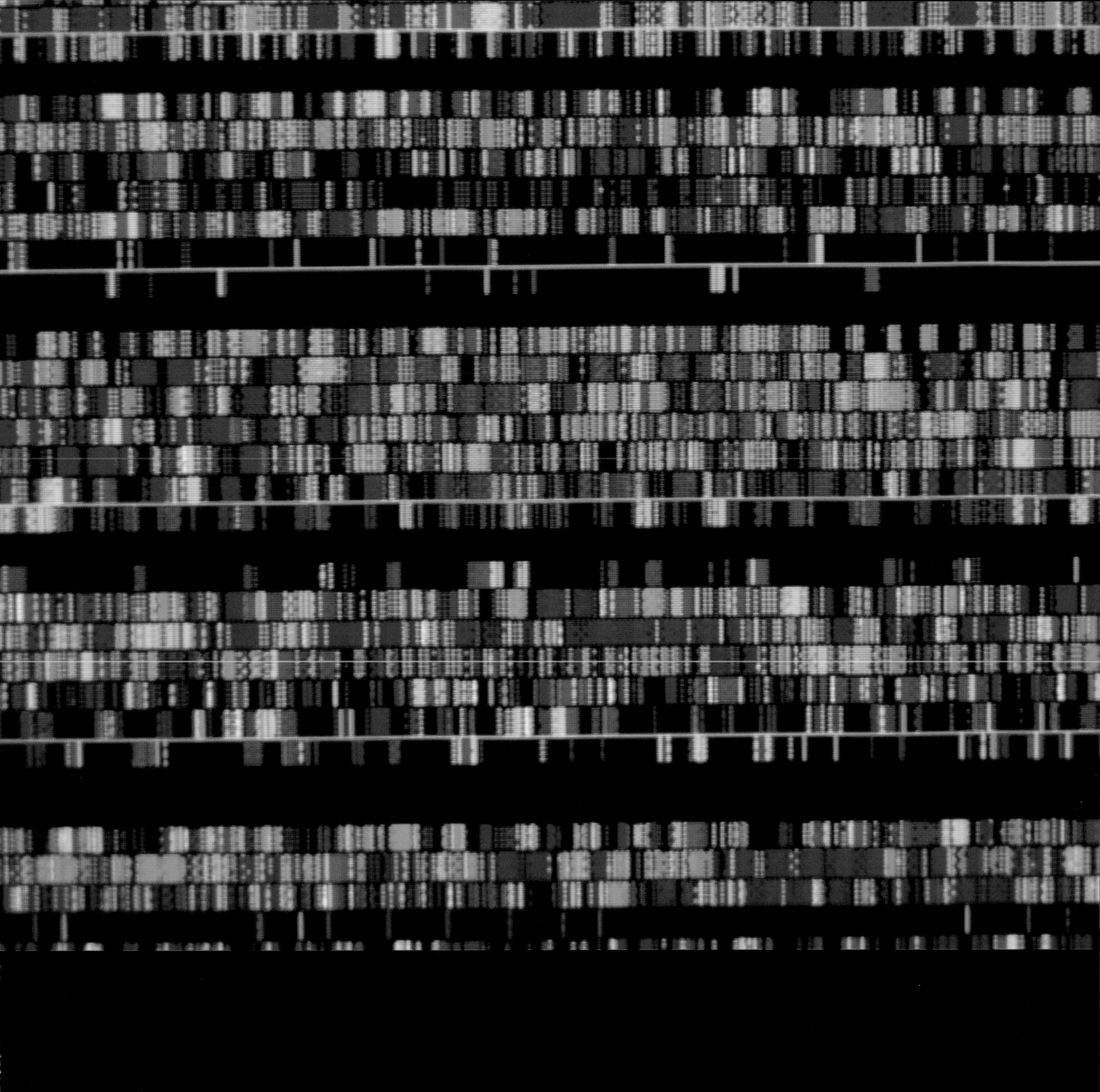

The transparent body
magnetic resonance imaging

Magnetic resonance imaging (MRI) is safer than CAT, since patients are not subjected to potentially harmful X-rays during their scans. Instead, a powerful magnetic field temporarily aligns all the atomic nuclei in the body in a particular direction, like trillions of little compass needles. While they remain aligned, a brief radio pulse is transmitted, causing the nuclei to go off alignment for a moment. When the radio pulse is switched off, the nuclei swivel back in line, still under the magnetic field's influence – and as they do so, they emit tiny radio pulses of their own. The MRI scanner then distinguishes the outgoing signals, which are characteristic for each type of tissue. These procedures are harmless, at least over the short time it takes to complete a scan.

The most extraordinary views of the human body's internal structures can thus be created, with colour-coding to distinguish different tissue densities. Computers store the scan data as raw numbers and values, and this allows researchers to 'tune' for the type of information they need.

The ideal medical scanner would eliminate the need for surgeons to make dangerous physical investigations of cancers, brain tumours, bone abnormalities and vascular disorders. Scanners are closing in on details a few millimetres across, but sometimes the potentially fatal flaws in a body are smaller than even our best machines can detect: pinprick clusters of cancerous cells, or ambiguous masses of tissue that may or may not need to be removed. Exploratory surgery is often performed to try and assess a patient's condition, and subsequently – perhaps days or weeks later – actual corrective surgery may take place. Every time the body is opened up and exposed to air, there is a risk of bacterial infection, even in the best-controlled operating theatres. Quite apart from that, patients undergoing surgery can suffer various forms of non-conscious bodily shock, unexpected reactions to anaesthetic, or heart palpitations and breathing problems. Some degree of blood loss is inevitable, and there is always a chance that surgical incisions will not heal properly when the body is closed up. Undoubtedly we will look forward to a new generation of scanners that will be sufficiently subtle in their investigations to make this potentially dangerous exploratory surgery redundant.

In the 1960s, futuristic films often showed medical scanners revealing the human body's deepest secrets at the flick of a switch. Modern medical technology is fast catching up with the predictions of science fiction, as shown by this MRI scan of a male subject.

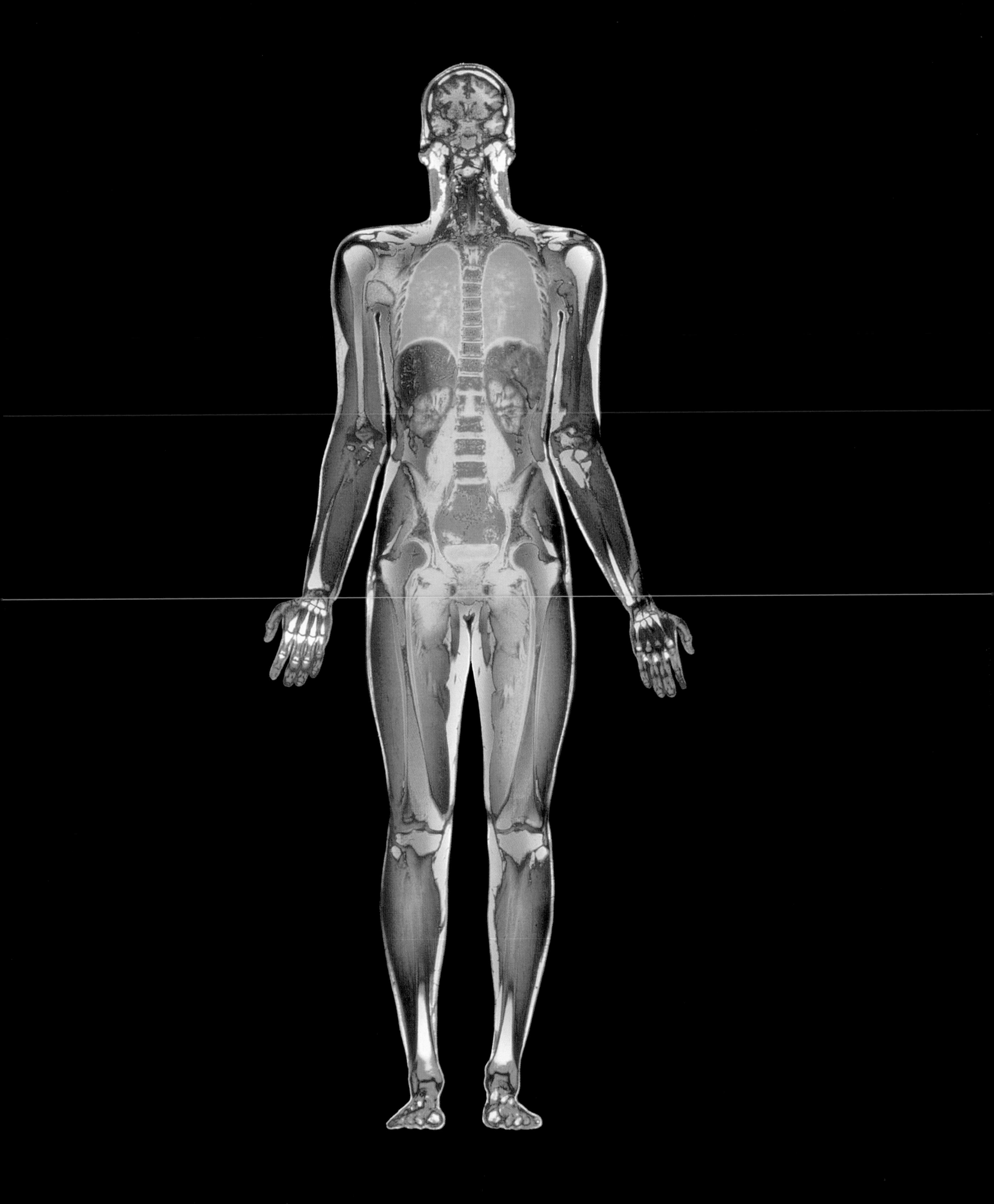

A rush of blood

magnetic resonance angiography

Magnetic resonance angiography is an adapted MRI technique that can be tuned to isolate blood vessels from the rest of the body. This is essential for detecting heart disorders, stroke damage and blood vessel diseases. A normal X-ray scan may not be able to distinguish blood vessels, because their tissues are usually too soft to block the rays. Angiography relies on special fluids, called 'contrast media', which are opaque to X-rays. Barium sulphate, a non-toxic compound of an otherwise extremely reactive poisonous metal, makes an excellent contrast medium. So-called 'barium meals' have long been used to highlight blockages in the digestive system, but these syrupy concoctions are not safe for injection into the bloodstream. In the 1920s, the Portuguese neurologist Egas Moniz injected iodine solutions into his patients' brains so that he could detect abnormalities and tumours. The iodine was semi-opaque to his X-ray machine, and the network of blood vessels inside the skull was highlighted in the image produced. Today a wide range of contrast media provides different flow rates, viscosities and degrees of opacity, and examiners can fine-tune their combinations of scanners and contrast media for the best results.

A modern MRA scanner can build a finely detailed map of blood vessels. The resulting pictures can be quite eerie: it is as though the bulk of the human body and its major organs have been rendered transparent as glass, so that only the blood system remains visible. Fine capillaries are still difficult to pick out, because of their microscopically tiny size and infinitesimal signal strength. If we could somehow stretch out every last blood vessel in an adult human, so that it formed a straight line, it would be 100,000 kilometres long.

In this MRI angiography scan of a human chest, the contrast medium has been selected to highlight a broad overview of major vessels and arteries in the heart–lung system. The aortic arch in the upper centre of the image carries oxygen-rich blood out of the heart. As the body burns energy, the oxygen is exhausted. Two pulmonary arteries (centre left and right) carry depleted blood back into the lungs. Pulmonary veins (lower left and right) then send re-oxygenated blood from the lungs to the heart. The bones in the ribcage were scanned separately with conventional X-rays and then combined with the MRI angiograph. It is not uncommon, today, for multiple scanner techniques to be incorporated into a single image.

An MRA scan of a chest cavity, combined with a separate X-ray image of the bones of the ribcage by computer processing. Multiple scanning techniques, often applied simultaneously, may be used to limit the time that a patient must spend undergoing tests.

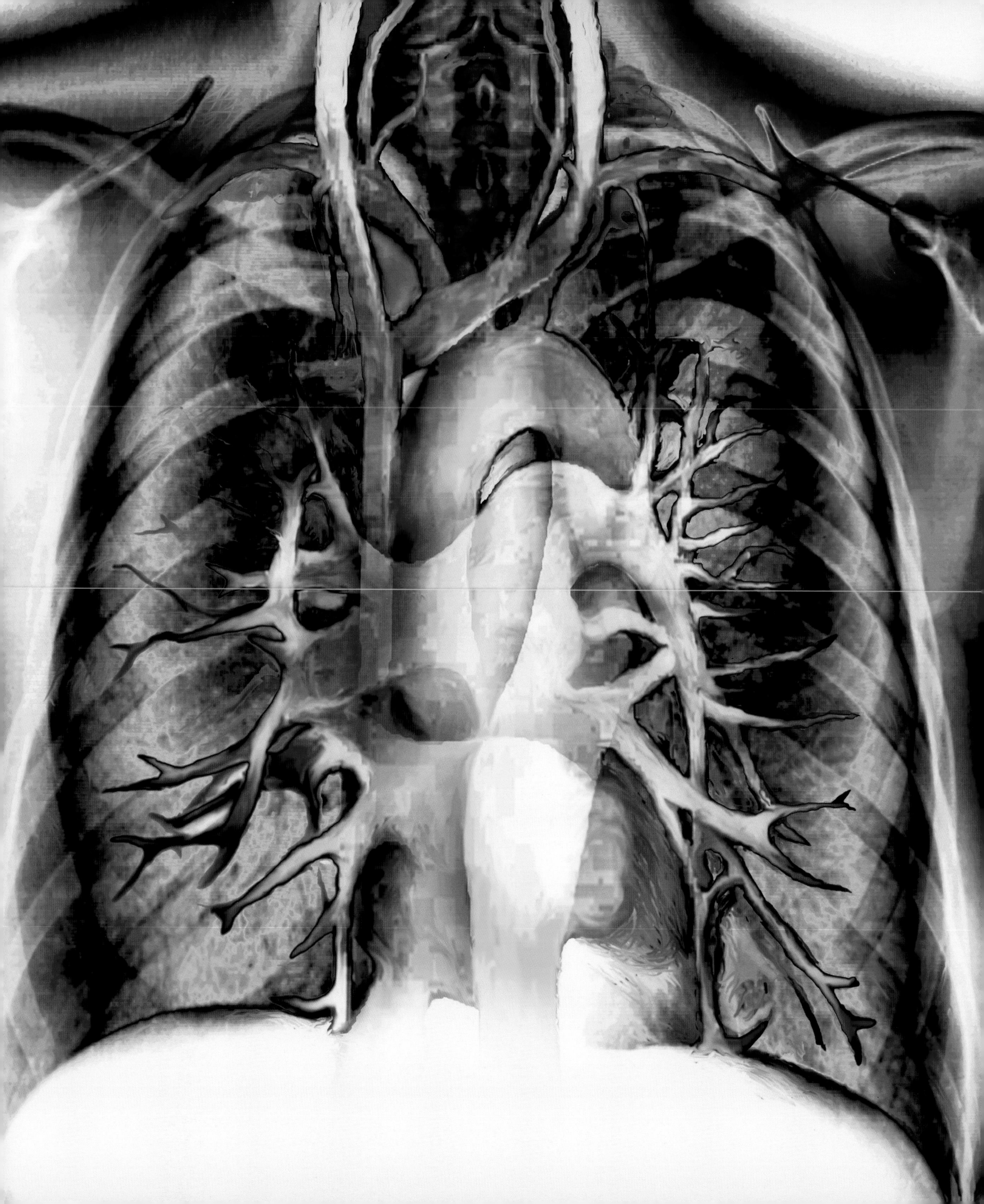

The brain at work

capturing a thought

'Functional' MRI allows actual biological processes to be observed, particularly in the brain. The magnetic properties of blood vary according to the amount of oxygen it carries and, just like muscles, brain cells need to consume oxygen when they are activated. Functional MRI highlights those areas of the brain that become activated during particular types of mental activity or emotional response. An even more advanced technique, positron emission tomography (PET), follows the progress of a weakly radioactive positron-emitting tracer chemical injected into the bloodstream. When the tracer reaches active areas of the brain, the positrons knock into nearby electrons, generating gamma rays that can be detected by the scanner. In this way we can observe areas of the brain that become activated during particular kinds of mental process: when subjects are agitated by distressing photographs, for instance, or calmed by pleasant ones.

The detailed maps of brain activity that have emerged from the new scanning technologies reinforce, and greatly add to, our existing knowledge of the brain's specialised regions of activity. The hippocampus, for instance, acts as a central processor for memories, storing and identifying different kinds of memory and eventually passing them to appropriate regions of the brain. A happy memory might consist of the sensory impression of a caress, details of a loved one's face, the colour of a dress, or factual information about the time and place of a magical moment. The overall memory of such a moment is, in fact, composed of many different kinds of sub-memory. When brain scanner subjects consciously remember, researchers can begin to glimpse where the various components are stored – or, more specifically, which parts of the brain 'light up' in response to memories being consciously retrieved.

The *conscious* experience of those memories, though, is not so easy to pin down in any locus of the physical brain. How do all the different areas connect to form consciousness? What special quality of the mind makes us a witness to our own thoughts? Brain scans reveal some of the physical mechanisms of thinking, but the awareness of our own existence is not something that we are able to explain – yet.

Multiple magnetic resonance scans of a brain show slice-like images which can be combined later to pinpoint features precisely in three-dimensional space. The image below shows a planned surgical procedure on the brain being mapped in advance.

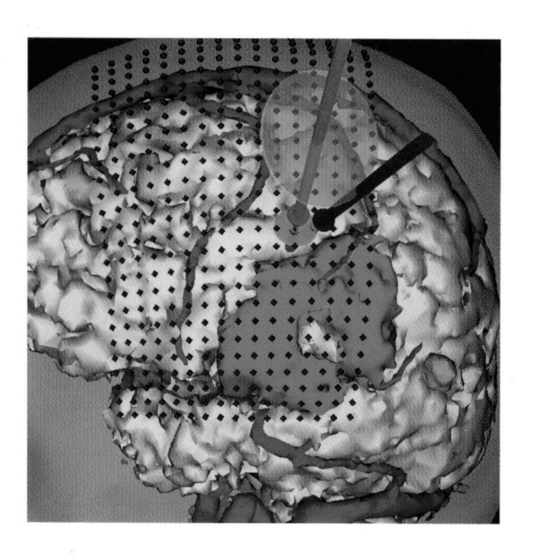

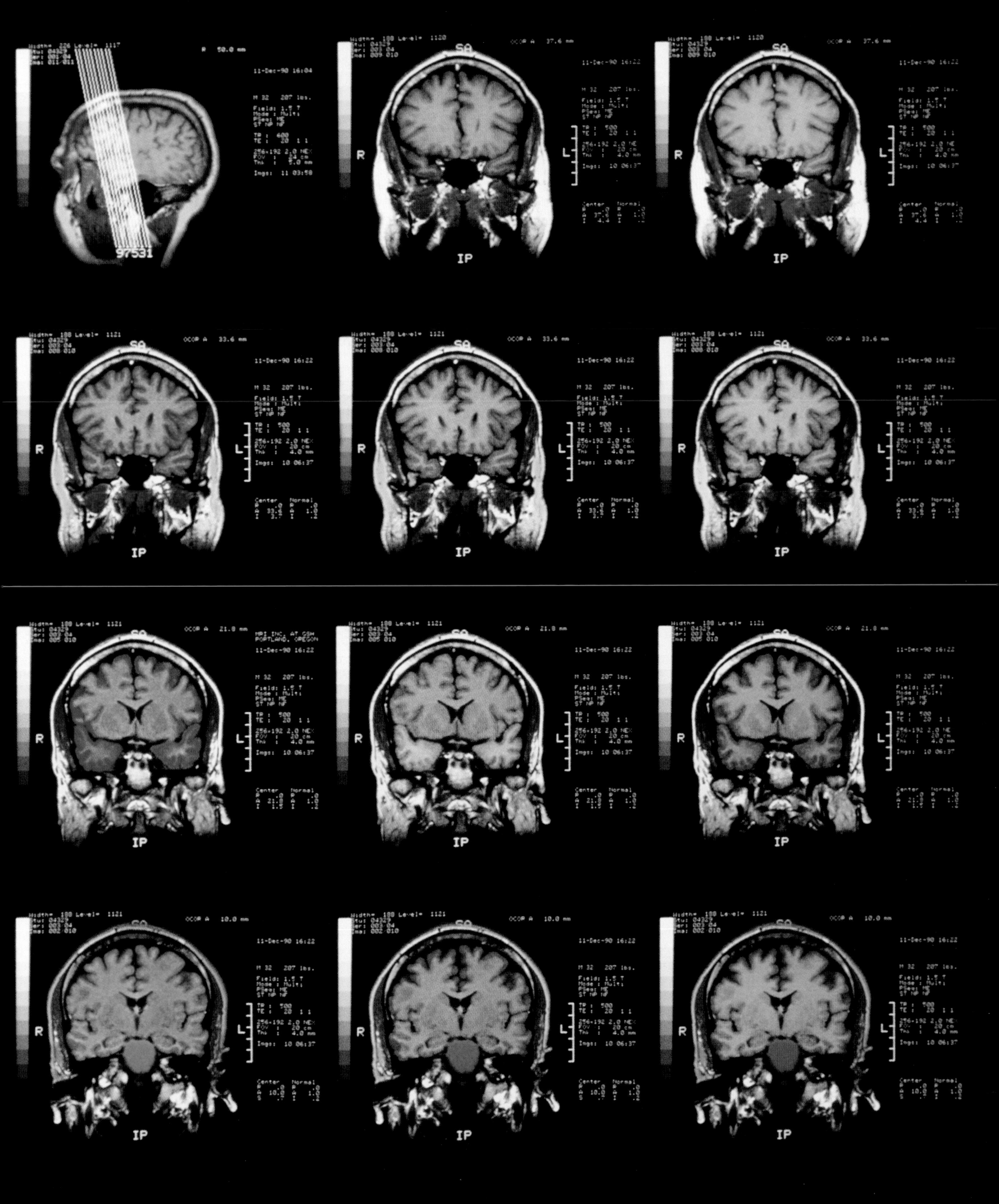

97531
11-Dec-90 16:04
SA
R
L
IP
11-Dec-90 16:22
MRI INC. AT GSM
PORTLAND, OREGON

Searching for the self
the brain's electrical field

For all their sophistication, most brain scanners are too slow to capture very much more than a rough impression of mental activity. Neurons (brain cells) communicate in millisecond intervals, but an imaging scan takes anywhere from several seconds to half a minute, depending on the specific technique. However, every electric current, no matter how tiny, must generate a magnetic field. The most expensive scanner of all, the magnetoencephalograph (MEG), is capable of detecting magnetic fields a billion-fold weaker than the Earth's natural field, generated by neurons as they fire. The room containing the scanner needs to be isolated from that field and the many other stray magnetic pulses generated by human electrical technology, and for this a costly shell of aluminium and nickel-iron alloys is required. MEG scanners can detect electrical activity in a single cubic millimetre of brain tissue, and record mental 'events' lasting just a few thousandths of a second.

Some researchers think that the secrets of consciousness will be discovered in the brain's overall electrical patterns, rather than in the physical structures of its nerve cells. Perhaps some kind of subatomic quantum field binds electrical signals from different regions of the brain into a cohesive whole? Or perhaps the waves of electrical activity that cascade through the brain at regular intervals help to coordinate its activity in a simpler way, just as a computer uses an internal 'clock speed' to run its high-speed calculations in a tidy order?

Consciousness, notoriously hard to define as a concept, is also extremely difficult to locate in the brain. It may be a property that emerges from the collective action of billions of unconscious brain cells, just as the apparently purposive and intelligent behaviour of an ant colony emerges from the collective reactions of individual ants who are not, themselves, particularly smart. But this may be nothing more than a mechanistic analogy. Some theorists argue that there is a profound difference between intelligent behaviour and self-awareness. The first, we might soon mimic in computer chips; the second, we have yet to fathom.

A microscopic image of brain cells, dyed to highlight details. The cells are made from chemicals, cell walls, axons, neurons, dendrites and other structures. We have yet to demonstrate how intangible qualities, such as consciousness and feeling, can arise from these materials.

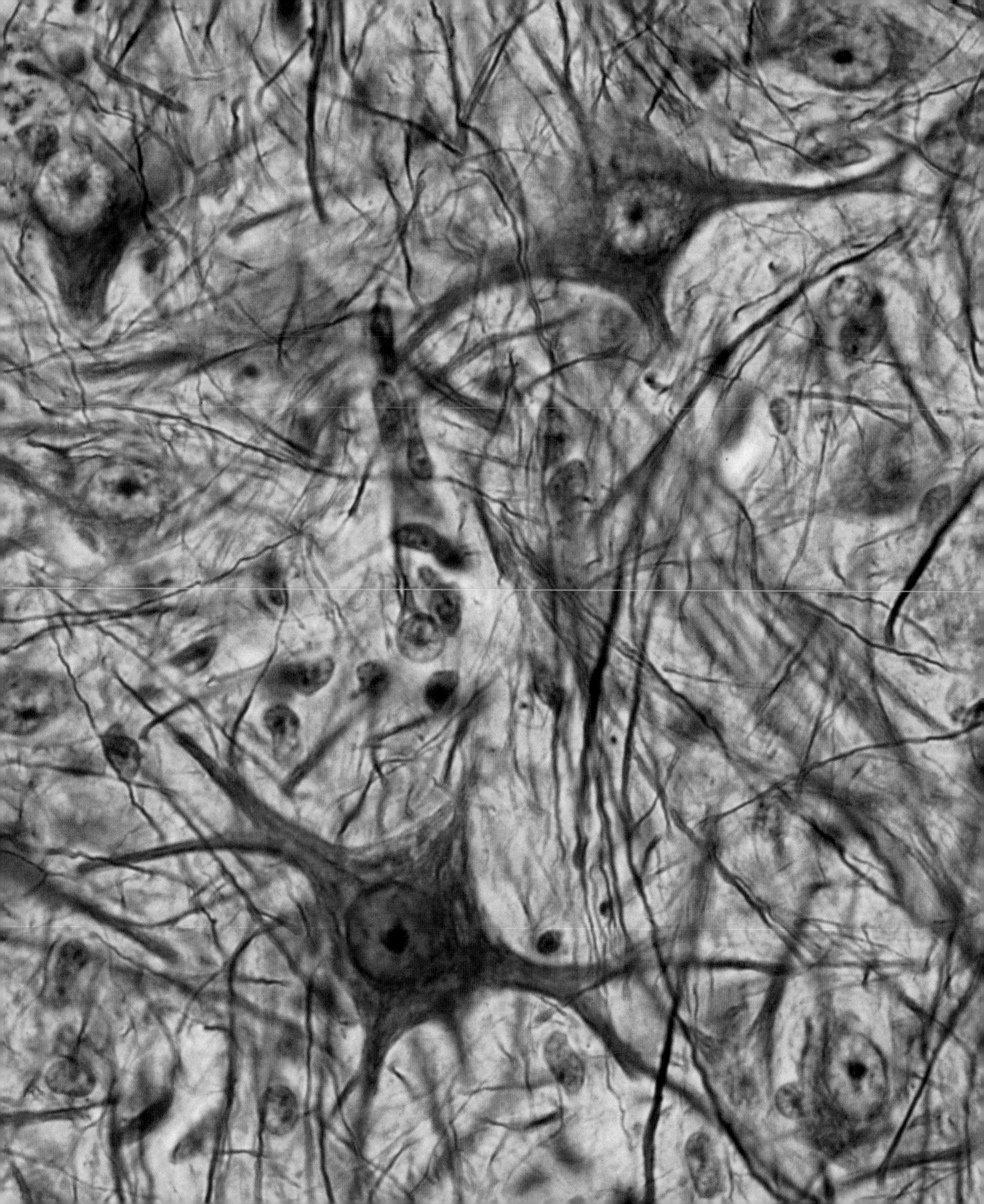

God in the head

images of transcendental states of mind

Scanners have revealed a correlation between specific types of brain activity and transcendental meditative states. Areas of the frontal cortex associated with concentration (situated in the forehead) become richly oxygenated by enhanced blood flow, but just as significantly, a part of the parietal lobe (upper middle of the brain) responsible for body image mapping is almost entirely shut down. Any mammal needs to know, from moment to moment, where its body ends and the outside world begins. In waking life, this mapping area of the parietal lobe is strongly activated, but in Tibetan monks calling upon years of meditative skill, this normally restless region becomes quiescent. The hyper-conscious frontal cortex focuses on an unusual (and arguably erroneous) message from the parietal lobe, resulting from its temporary failure to keep the frontal cortex aware of sensations from the outer extremities of the body: 'The bounds of your body are blurred, and the outside world is part of you.' Is it possible that the subjective 'one-ness with the universe' feeling so often ascribed to transcendental experience is actually *caused* by this parietal shutdown? Is spirituality nothing more than a coldly quantifiable anomaly in the brain?

Researchers in this field are not necessarily claiming that spiritual states are merely a delusion of the mind. They are saying something much less obvious. Science is all about what can be weighed, counted, measured. Whether or not God exists is not a question that neuroscience alone can answer. However, mystical experience is a commonplace biological phenomenon that is now scientifically observable. Neuroscientists are not trying to start a religious war; they are simply concerned to discover *why* the capacity for altered states of consciousness seems to have been hard-wired by evolution into our brains.

This scan maps the temporal lobe region in the brain of a schizophrenic patient during a severe hallucinogenic episode. Schizophrenics may report hearing voices in their heads, and they sometimes describe the feeling that God has communicated with them. A normal brain would deliver a much more symmetrical pattern of activity in the left and right regions. In this case, the applied imaging technique was positron emission tomography (PET), in which a mildly radioactive tracer is injected into the brain's bloodstream. The tracer accumulates in brain areas that are particularly active.

A hallucinatory schizophrenic episode captured in a PET map of heightened activity in the brain. PET scans require a mildly radioactive tracer to be injected into the bloodstream of the brain. The tracer attaches itself briefly to areas where metabolic activity is heightened.

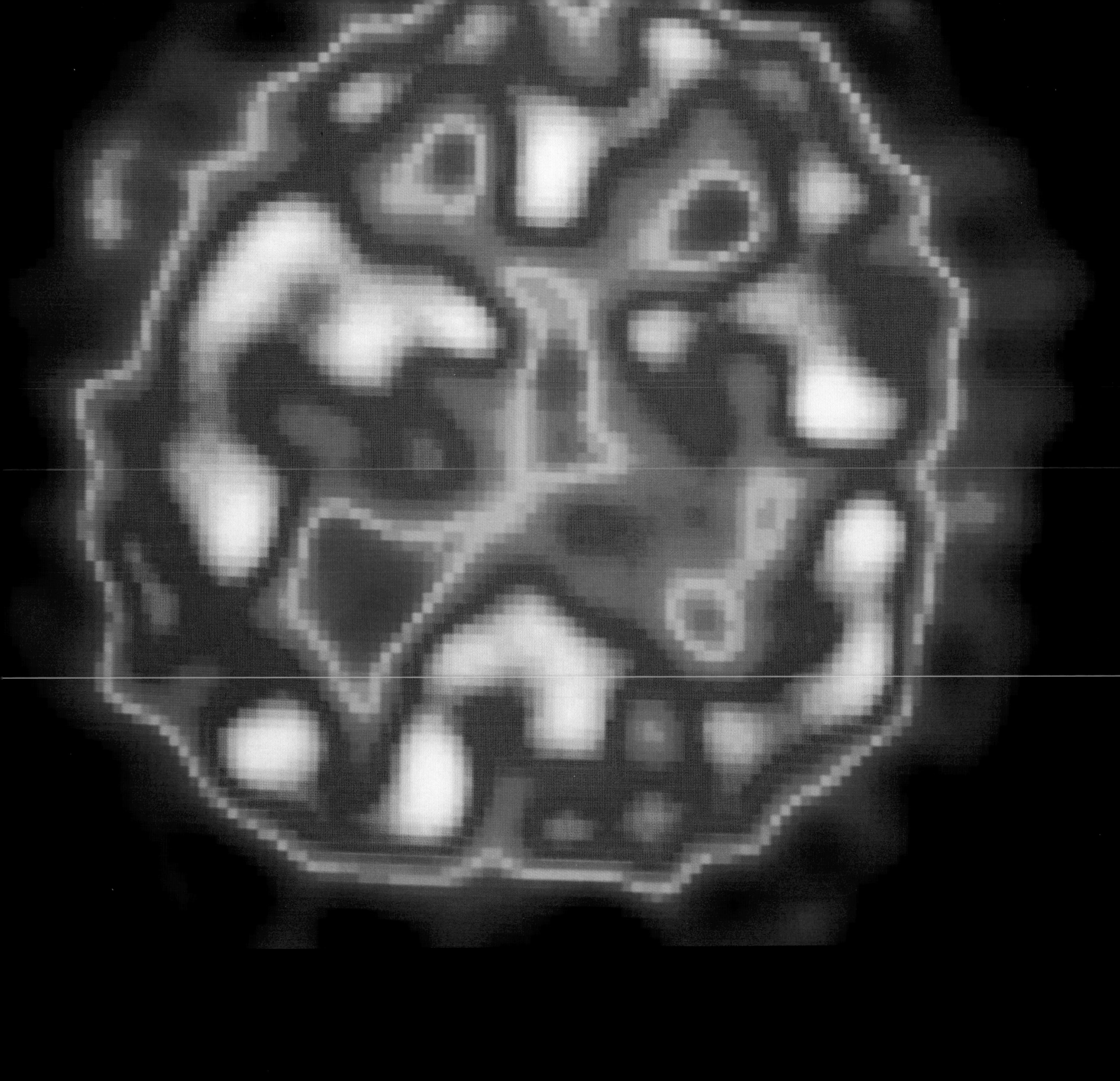

The digital human

a multilayered model of the entire body

In August 1991, the US National Library of Medicine (NLM) commissioned an extraordinary project: the complete mapping of a human body, based on an array of CAT, MRI and other scans to be conducted at the University of Colorado. Nearly 2,000 cross-sectional scans were made of an unnamed 39-year-old convicted murderer who had agreed to submit his body to science after his execution. Once initial scans of the fresh cadaver were completed, it was embedded in gelatin, frozen, and sliced from head to toe. As each layer was exposed, a colour photograph was taken and digitally scanned. Today, a wide range of researchers can gain access to the so-called Visible Man, either in the form of images, or as unprocessed computer output from the scanners.

The taboo against desecrating the bodies of the dead goes back many centuries. It was prohibited by ancient Greek and Roman religions. The Renaissance saw a surge in the study of anatomy, driven by the curiosity of artists such as Leonardo da Vinci, who may have conducted as many as 30 dissections in his lifetime. In 1510, he recorded the similarity between human and animal muscle structures – but dared not publish his work. In 1565, London's Royal College of Physicians authorised human dissections, and by the eighteenth century the bodies of condemned criminals were being purloined for medicine – yet still there were not enough to go around, and some surgeons resorted to grave-robbing. In 1829, the hunger for dissection created its inevitable extreme, when the notorious William Burke was hanged in Edinburgh for suffocating his victims and selling their bodies to physicians.

The Visible Man project fits neatly into this tradition, with its mix of criminality, taboo and dedicated artistry. The use of computers to make pictures occasionally seems to take away human skill and replace it with an unimaginative, mechanical vision. In this case, however, the NLM has provided swathes of data that can be interpreted and assembled in many different ways. The art lies not so much in the wielding of a pen and paper, but in the ingenuity of the software programs that are written – by imaginative humans – to reassemble Visible Man in new ways. Some of the results, as in this example from a team at the University of Hamburg, would surely have met with Leonardo's approval.

Detailed MRI, CAT and other scans of a male body have been distributed to medical researchers around the world by the US National Library of Medicine. This image was created at the University of Hamburg, but it is only one particular interpretation of the data.

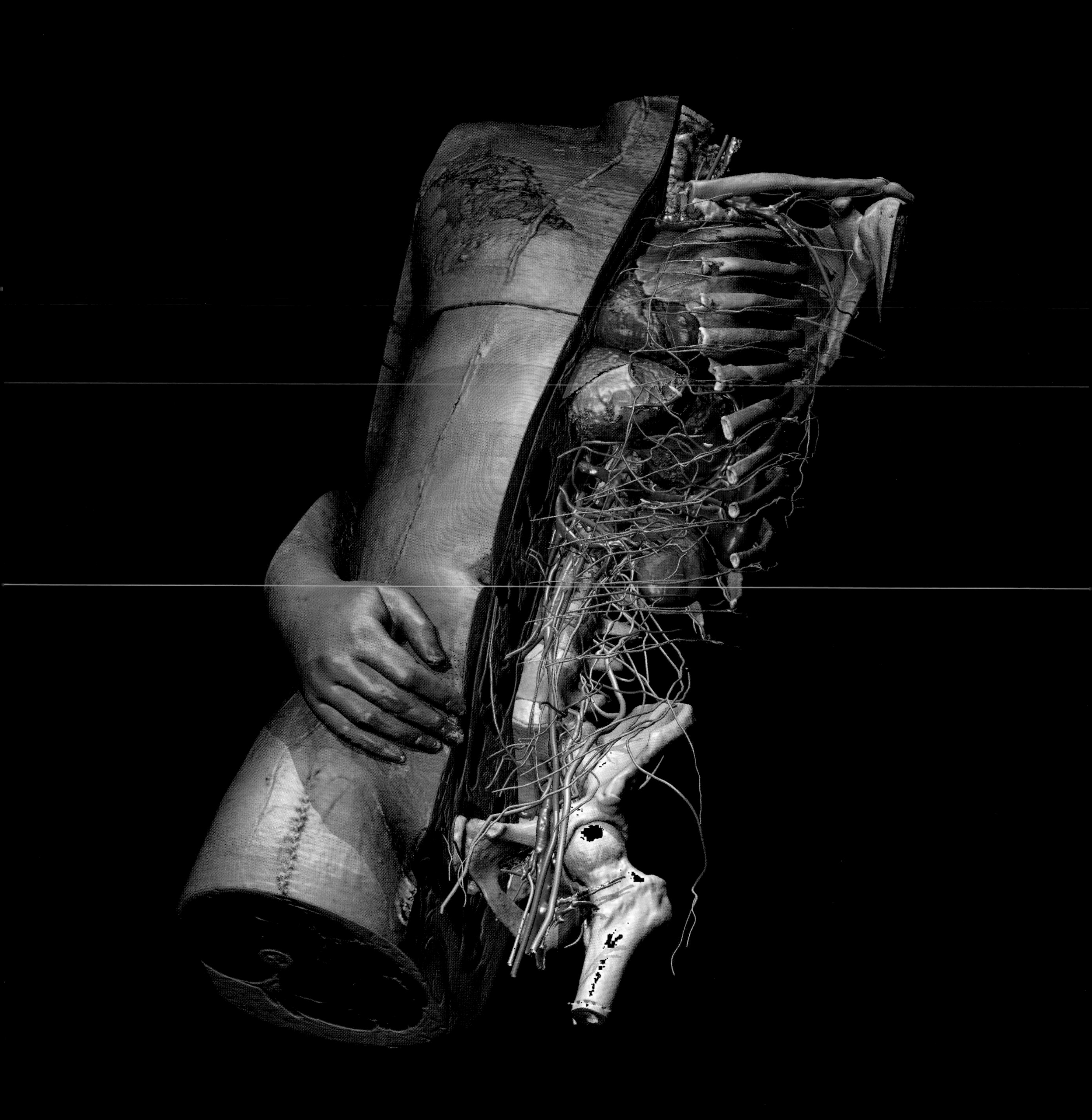

The ultimate scanner
exploiting the 'terahertz gap'

At the point of the electromagnetic spectrum where light and radio waves merge, the 'terahertz gap' awaits our exploitation ('tera' means 'trillion'). At the 10 THz end of the terahertz spectrum, just below the region exploited by infra-red remote control television handsets, the waves are referred to as 'far infrared' and behave more or less like light. At the lower (0.1 THz) end of the gap, just above the microwave zone used for mobile phones, the waves are known as 'millimetre waves' and behave more or less like radio. Obviously some combin-ation of light and radio imaging would be a desirable technique. Radio signals can pass through a wide variety of materials that block light, while light rays can be easily focused into a fine image, yet are blocked by the thinnest sheet of paper. Imagine a crisp photograph of an object hidden behind a wall, and this gives some idea why researchers are excited by terahertz radiation.

Until recently, development of terahertz technology was hampered by the lack of a convenient source of radiation to transmit the waves, and sufficiently sensitive receivers to detect them. Early experiments relied on cumbersome equipment that only operated at liquid helium temperatures and required the presence of a strong magnetic field; and the detectors required aerials so finely engineered that no profitable commercial uses could be found for them. Now, extremely fast-cycling semiconductor circuits and compact lasers allow terahertz signals to be produced and detected at room temperature, from devices small enough to fit in a briefcase.

In the spring of 2003, a European Space Agency team released a terahertz image of a human hand taken through a pad of paper; meanwhile, library researchers were exploring how to read the supposedly irretrievable texts of ancient volumes whose pages had become stuck together because of damp or decay; and manufacturers in Britain and the United States answered growing concerns about airport security in the wake of the Twin Towers catastrophe by unveiling machines that can detect weapons hidden under a passenger's clothes. The results are so revealing that they have worried civil rights and privacy campaigners, who argue that the 'T-ray' scanners reveal too much detail of a person's intimate anatomy. Scanners under development for practical deploy-ment in airports will need to replace the actual body outline with anonymous computer simulations in order not to cause offence.

Terahertz scanners are in the earliest stages of development. Government agencies and security companies are unwilling to reveal too much about their current uses or image resolution capabilities. The picture on the right, made with existing technology, gives some flavour of what a terahertz-derived image might look like in the future. A man, stripped to the waist, has been scanned by laser, and his exact bodily measurements have been calculated. The procedure has lasted barely three seconds. The purpose? Bespoke tailoring of fine quality gentlemen's suits. Terahertz devices may obtain similarly distinct personal data about us without requiring that we remove our clothes – or that we are even aware that a scan is taking place.

The so-called 'holographic tailor' uses laser light to scan a male torso, providing exact measurements of the body. Future security scanners will search for unique identifying markers, such as genetic information, facial features or scans of the retina.

4

Supersenses

Some technologies blur the boundaries between pure scientific research and its everyday practical applications. We welcome sophisticated hospital scanners, but are much more wary of devices whose purpose is to keep track of us as we go about our daily lives. There is no easy boundary to be drawn between one human merely looking at another with the naked eye, and machines that allow unprecedented observation.

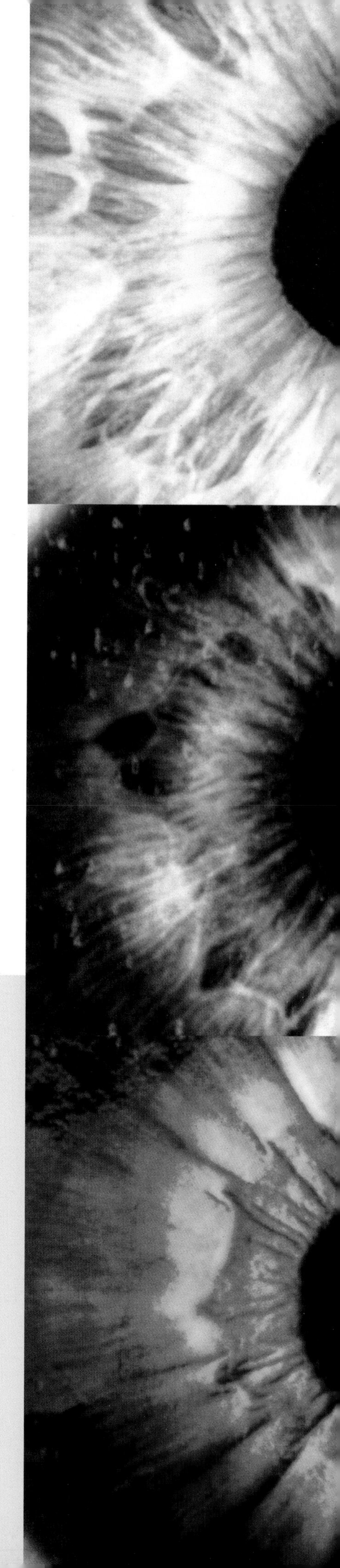

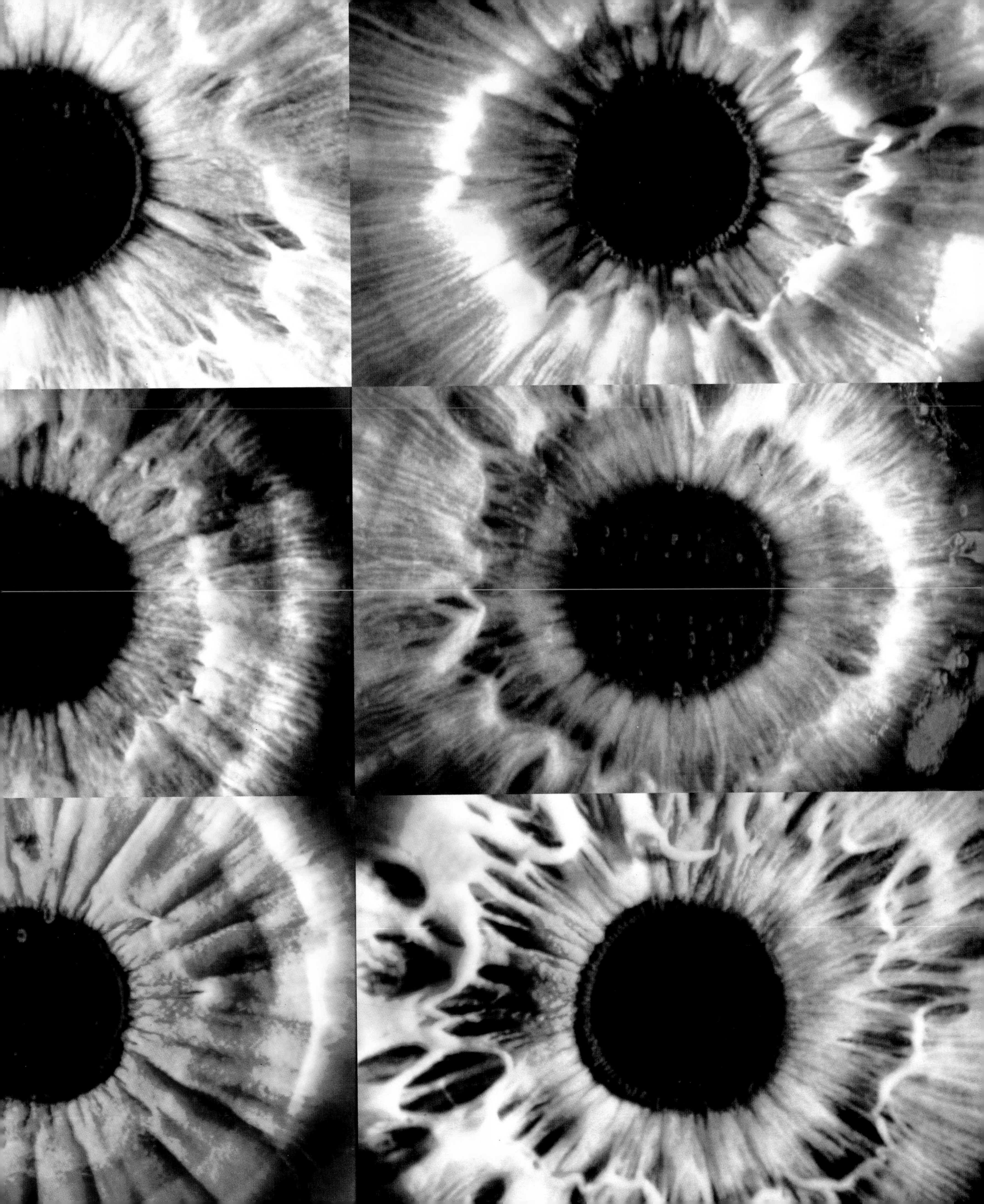

What we do see
the human eye's abilities

Soldiers are used to handling equipment that enhances visual acuity in certain areas, night-sights and infrared scopes in particular. However, they also value the abilities of a standard-issue device they call the 'Mk. 1 Human Eyeball'. Much of this book focuses on what the human eye cannot see, yet it is quite remarkable what we *can* see. Our eyes can focus on objects anywhere from 15 centimetres away to the far horizon. They can adapt well to low-light conditions: after 20 minutes' exposure to a dark night, we can see the tiny pinprick of light from a single candle three kilometres distant – it takes just six photons to trigger a response in a retinal cell. The central portion of the eye can resolve details as small as one-tenth of a millimetre at arm's length. On a bright day, we can discriminate between 200 colour hues, 20 saturation values and 500 different levels of brightness: in all, this amounts to approximately 2 million colours.

Other animals do better in certain areas. Kestrels and other raptors have a fantastic ability to spot the tiniest movements of prey from a great altitude, and many non-predatory birds have a far wider perception of colour than we do. But on balance, our eyes are good, especially when allied to the perceptual systems of the brain. We have an unmatched ability to discern patterns, to recognise shapes and objects, by comparing raw input from the eye with the brain's vast reservoir of remembered patterns and associations. 'Seeing' happens more in the brain than in the eyeball.

It might have been evolutionarily useful for us to be able to see in infrared and ultraviolet, as well as the colours that we already see; but nature has compromised, limiting our vision to a group of wavelengths that are close enough together to be neatly brought to a focus by a single lens. If the extremes of our wavelength range were too far apart, no one lens could bring them all to the same focus, and our vision would be blurred by 'chromatic aberration', the smudgy rainbow effect we get when looking at the world through a plastic lens of poor quality. The optical region of the electromagnetic spectrum is nothing special; it just happens to be a sufficiently narrow range for our eyes to cope with efficiently.

PRECEDING PAGES
The human eye was once considered to be the result of 'perfect' design at the hands of a divine creator. But there is no such thing as a perfect eye. Some animals have eyes that are much better than ours at detecting movement, and others can see colours inaccessible to us.

FACING PAGE
An electron scan of a portion of the iris in a human eyeball, which adjusts the amount of light that is allowed to pass through the eye and hit the retina. On a dark night, the iris opens wide to allow in as much light as possible, but bright sunlight provokes the opposite reaction.

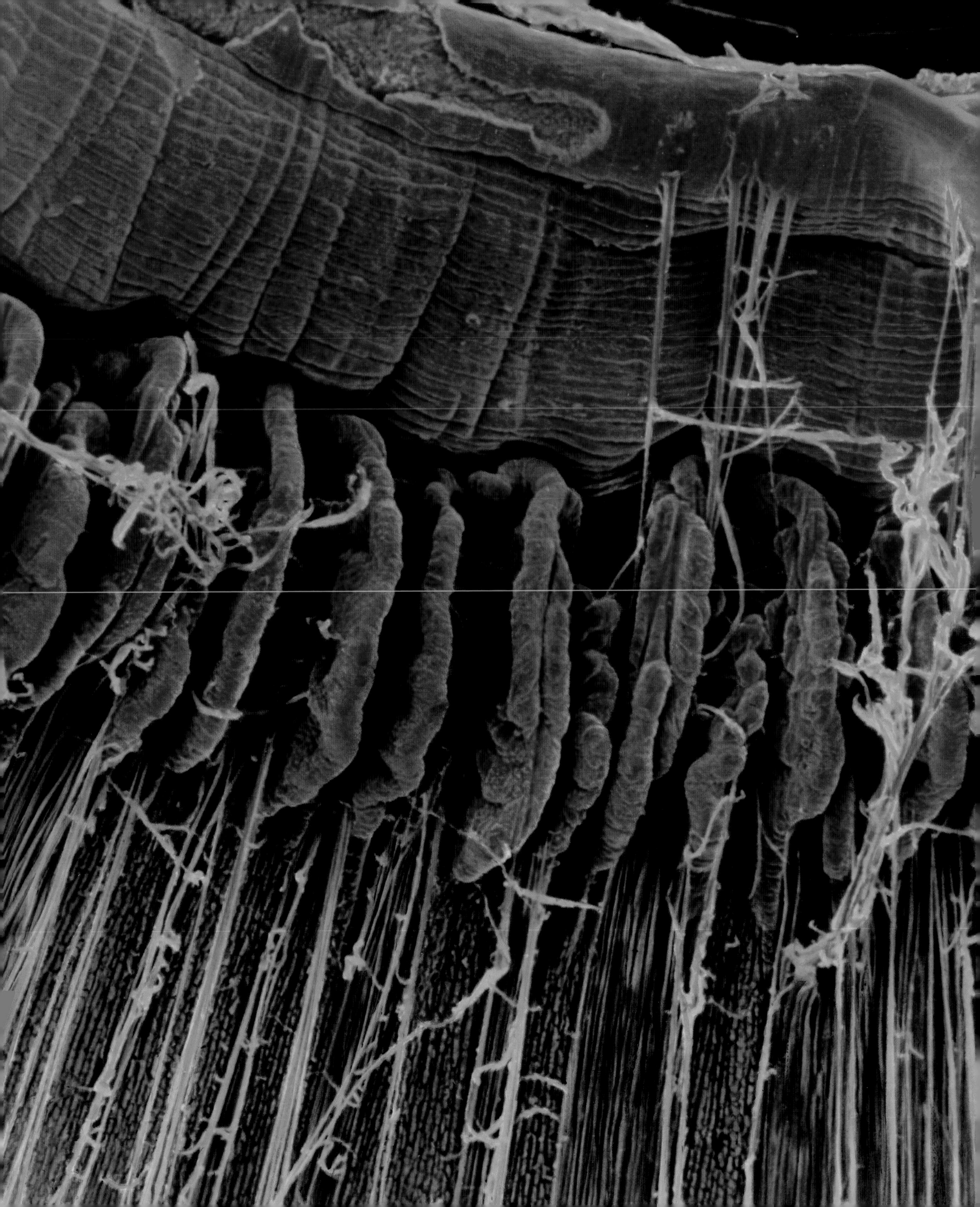

Being a bee
insects & ultraviolet vision

Bees cannot distinguish the red end of the spectrum, which they perceive as black: an absence of light. They can see green, yellow, purple, blue, and beyond into the ultraviolet frequencies just outside the limits of normal human vision. Many pollen-bearing flowers have markings that reflect or absorb ultraviolet in distinct patterns. The prettiness of a flower's colouring, as appreciated by us, is irrelevant to a bee.

Why is ultraviolet crucial to flowers and bees, but not to us? Part of the answer lies in the way a bee navigates, using polarised sunlight. Even if the Sun is obscured by clouds, some ultraviolet still gets through, because its energetic wavelengths penetrate the cloud layer. The bee, therefore, is not reliant on clear skies to navigate. Flowers need to advertise their nectar to passing ultraviolet-sensitive insects, and so at least a third of all pollen-bearing flowers include dark (ultraviolet-absorbing) markings with no obvious equivalents in the visible spectrum. A yellow marigold, for instance, seems uniform to us in colour, but a bee can zero in on what it perceives as the 'dark' bullseye of its nectar and pollen organs.

There are subtle reasons why humans are excluded from a bee's-eye view of the world. Ultraviolet is harmful to tissues – hence all the sun creams we use to protect our skin from it. A bee can tolerate damage over its short lifespan, but a person cannot; and so the lenses in our eyes block ultraviolet, protecting the retinas against sunburn, but at the same time preventing us from seeing it. However, something interesting happens to cataract sufferers when their clouded lenses are surgically removed. They can see ultraviolet as a kind of blue, because there is no longer a lens in the way. Military intelligence exploited this in the 1940s, recruiting ultraviolet-sensitive observers to watch German U-boats signalling with ultraviolet lamps; and Claude Monet's use of colour changed after his cataract surgery in 1923. He painted waterlilies bluer than before.

We can scarcely imagine how a bee, with its 6,900 compound lenses in each eye, might actually 'perceive' a flower. But we can at least distinguish a flower's ultraviolet markings using photographic techniques that block out visible light and allow only ultraviolet frequencies to pass into the camera's imaging system.

An ultraviolet light photograph of a daisy with nectar guides on its petals. In ordinary daylight, the guides are invisible to human eyes, and the petals appear uniformly white. The compound eyes (below) of bees and many other insects are sensitive to ultraviolet.

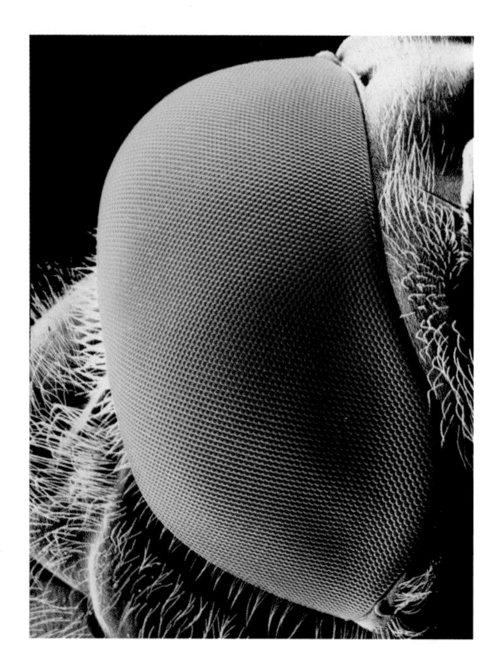

The revealing glow
ultraviolet's forensic uses in criminology

Criminals cannot help but leave evidence behind them. The human body incessantly sheds skin and hair, and the palms and fingers exude sticky oils which add extra grip but which also leave traces on any reasonably smooth surface. Fingerprints can be detected and preserved merely by sprinkling a fine, light powder over the contact area, and picking up the resulting pattern with sticky tape. But where to look for prints in the first place, when they are not immediately visible to the naked eye?

An ultraviolet lamp reveals organic traces at a crime scene, in just the same way as fluorescence microscopy uses a higher-energy light to cause organic substances to glow at a lower frequency. Fluid dyes that are transparent in daylight are lightly sprayed over suspect areas. Dyes in the spray attach themselves to any organic matter they find, including the oils left in a fingerprint. Under ultraviolet lighting (which itself remains invisible to human eyes) the dyes glow bright blue or green, revealing the fingerprints, or other bodily traces left at a scene. The tiniest speck of blood on the floor of a car, or in a bedroom carpet, can be located and retrieved for analysis. Why, then, should any defence lawyer go to the trouble of arguing against such seemingly conclusive scientific evidence in court?

Forensic science neatly encompasses the great problem of any instrumental observation of the world. The ultraviolet lamps and other detection devices show that traces of organic material exist at a scene, but they do not necessarily reveal how they got there, or what they mean. A lawyer may argue that the forensic teams themselves could inadvertently (or *perhaps* deliberately) have polluted a crime scene with their own or someone else's bodily traces. Specks of blood can be ejected in an innocent sneeze, or deposited on a carpet by a stray footfall. Skin cells leak from even the best forensic body suits. Fingerprints can be planted in compromising places by reversing the sticky-tape collection procedure.

Forensics is considered a science because the profession requires great rigour in its gathering and identification of evidential traces; however, just like any other branch of science, it is not always a clear-cut question of absolute truth or falsity, as we might like to think. The evidence-gathering techniques deliver specific pieces of information, but then it is all down to how people choose to interpret the data. The theory that best explains the evidence may win the day in court – but not always.

A fingerprint, stored as a digital file so that it can be more easily cross-referenced with other prints. The exact arrangement of whorls in a print is unique to an individual, but the general patterns are sufficiently similar that careful analysis must be made for identification.

Pseudo-science? Kirlian photography & the paranormal

Could there be any scientific basis to paranormal phenomena such as spirits, psychic energy or ghosts? One imaging technique seems to show there might be – depending on how we interpret the results.

In 1961, Semyon Kirlian and Valentina Kirliana published a paper in the *Russian Journal of Scientific and Applied Photography*, describing what is now known as Kirlian photography. The process is fairly simple. A sheet of film is laid on a metal plate, and an organic object is placed on top. A high voltage is then applied to the plate to make an exposure. After processing, the film shows the shadow of the object, surrounded by a glowing haze. Many practitioners believe Kirlian photography reveals an invisible 'aura' that surrounds all living things: some kind of bio-electric energy field. It is fair to say that Kirlian photography is not yet fully understood in all its details, but the forces involved are almost certainly amenable to scientific rather than paranormal explanations.

Kirlian photography exploits something akin to static electricity. Some atoms hold on to their electrons more tightly than others: organic compounds cause our hair, skin and clothing to give up electrons more easily than, say, brass door handles. After a brisk walk along a carpeted corridor on a dry afternoon, we lose electrons and our bodies become positively charged. A difference in electric potential between the body and the awaiting door handle has to be resolved, and an impressive spark bridges the gap while our finger is still some way short of actually touching the handle. Kirlian photography sets up a similar difference between the charged metal plate and the subject being photographed. Care has to be taken so as not to electrocute living subjects, and there are various ways of setting up the apparatus, but the net result is that a mild sparking discharge leaps between the subject and the plate, passing through and exposing the photographic film on the way.

Some exponents claim that damaged leaves produce different auras than whole ones, and happy people yield better auras than depressives. However, this is a matter of conjecture. From a scientific point of view, the problem is that Kirlian results cannot be calibrated consistently. The technique is recording something, but exactly how useful or relevant it is no one can be sure. Sceptics say that moisture on the subject's surface is the main factor affecting the strength or weakness of an aura: tense people sweat, so they have damper skin, which weakens their electric potential, thus reducing the sparking phenomena. Perhaps this explains why a nervous person's aura seems weak?

Kirlian photography of a healthy cannabis leaf (*cannabis sativa*) shows what some might consider a uniform, healthy-looking aura. An experienced transcendental meditator (below) seems to give off sparks of energy when photographed by the same technique.

Hot pursuit
thermal detection of humans

Once a police helicopter has located a criminal suspect, it is all but impossible for him to hide. Let's call him Bad Guy. Where can he go? Into a building? It will simply be surrounded by police officers on the ground, and searched room by room until he is cornered. If Bad Guy steals a car, he might enjoy the illusion of a fast getaway for a few moments – but a car can only go along roads, it cannot dash down subway steps or into buildings. Besides, a car is easy to follow from the air. Maybe Bad Guy can skip the car idea and hide in long grass, or tall crops of wheat, or seek cover in a dense forest? Again, this is unlikely to help his escape. Even the darkest nights can no longer protect him. Thermal imaging captures the upper portion of the infrared light spectrum, which is emitted as heat by all objects, and it is routinely used to pick up even the faintest human heat signature against the colder background of the landscape.

Bad Guy's best chance of escape is to find his way into a crowd. Perhaps he will come across revellers streaming out of a pop concert late at night. Maybe he is clever enough to realise that his own heat signature will become almost indistinguishable from theirs if he mingles with them.

For reasons of cost and ease of use, most thermal cameras used for police work in recent times have delivered simple black-and-white displays of heat, but machines are now emerging that can colour-code the subtlest temperature differences. Bad Guy's body may be warm after his exertions running away from the helicopter, and the concert-goers might be pretty hot too, after dancing the night away – but there is a crucial difference. He has been generating all his heat from the inside, while the pop fans have been warmed from the outside too, by each other and by central heating in the building. The outer layers of Bad Guy's clothes will not be as warm as those of the people around him. It will not be long before computers and sensors can extract this kind of information, and Bad Guy will have nowhere to hide.

Thermal imaging cameras are routinely used on police helicopters and coastguard cutters to hunt for fugitives who may think they cannot be seen in the dark. Similar cameras attached to telephoto lenses are used by military forces to aim long-distance weapons at night.

Going with the flow
the hidden dynamics of air

We think of air as transparent, invisible, featureless; but a process known as schlieren photography can reveal its hidden dynamics. The techniques are varied and often complicated, but the underlying principles become clear to us on a warm summer's day, when a flat road surface shimmers in the heat. The air itself seems to ripple and glisten. However, it is our perception of what we see through the air – the far horizon and sky, perhaps – that creates this impression. Light is refracted haphazardly by layers of different densities of air as it tumbles and swirls under convection, heated from below by the hot tarmac. Alternatively, think of a candle sitting on a shelf in the middle of a darkened room. When a powerful, focused torch beam casts a shadow of the candle and its flame on the far wall, variations in air density above the flame can be seen as shimmers in the shadow. This is essentially the effect exploited by schlieren photography to capture the patterns of air flow and shock waves around wind tunnel models of aircraft, missiles and other high-speed aerodynamic machinery. By placing the camera, the test model and a strong light source along nearly the same line, the different densities of air around the model can be depicted.

A wind tunnel model is fixed, and its surrounding shock wave, generated by a powerful fan (or even a jet engine) at the end of the tunnel, is relatively stable and can be captured in a single image. Schlieren images of real aircraft speeding across the sky are more of a challenge, because both the plane and its shock wave are in rapid motion. Precision flying and careful planning are required to ensure a precise alignment of the Sun (the strong light source), the plane and a camera on the ground. A single photographic exposure under these circumstances would reveal little more than a flat blue sky behind the plane. However, a shot taken a fraction of a second later might show the tone of blue in the background sky to be slightly different, because changes in the density of the air around the passing aircraft would have caused the light to be slightly refracted before reaching the camera.

The schlieren technique, used to make the image on the opposite page, relies essentially on capturing those variations over a span of time (albeit less than a second) in many sequential shots, yet all on the same piece of film. A narrow vertical slit moves from left to right across the film, keeping pace with the movement of the jet across the field of view. Moment by moment, the variations in light levels are recorded as parallel strips accumulating along the film. The camera mechanism is geared to move smoothly and fluidly, so that the strips merge into each other to make what appears to be one seamless image.

The first schlieren photograph of a plane in flight: a NASA T-38 jet creates a supersonic shock wave high above the Edwards Air Force Base in California, from where Charles Yeager took off to break the sound barrier in 1947. The image below shows air and heat flow around a shuttle.

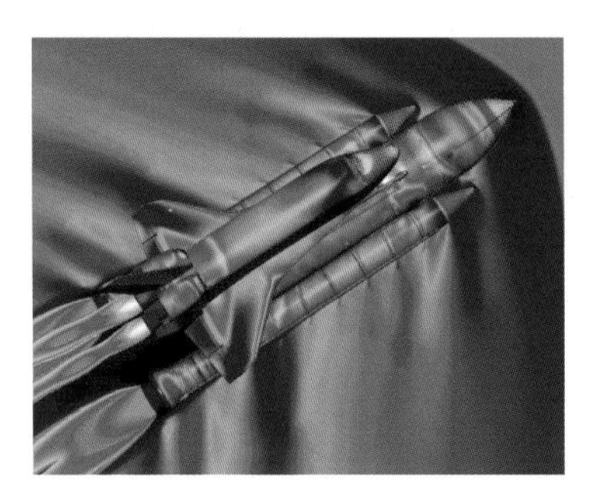

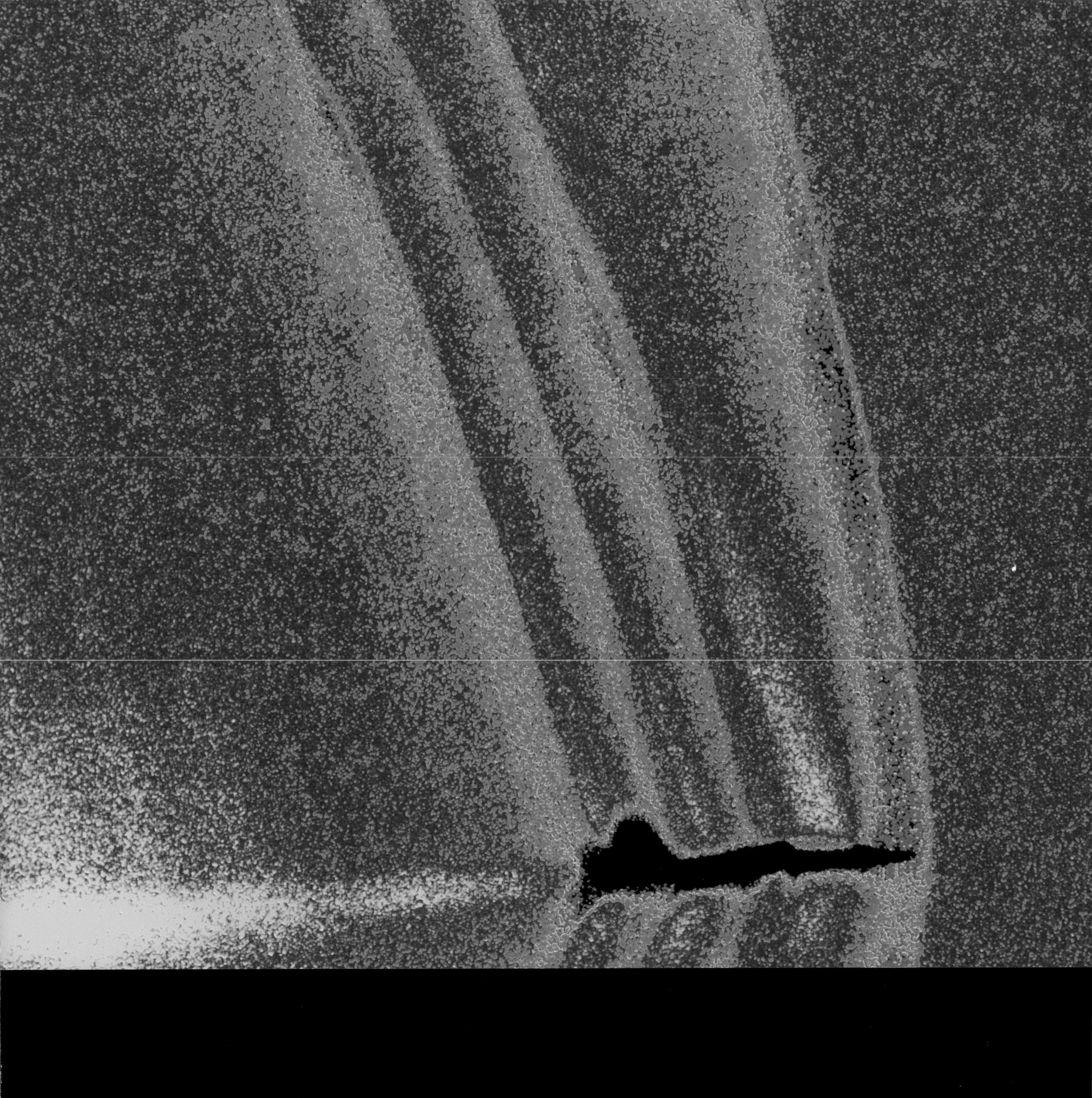

Under observation
the impossibility of privacy

Much of this book deals with our ability to detect invisible things. But a couple of everyday accessories in your pocket may be making you detectable to others, even at a great distance. Some mobile phone companies permit anxious parents to locate absent children to an accuracy of a few metres. The phone company's computers trace the map coordinates of the child's last phone call; the distances between the phone and nearby radio relay masts can be calculated by triangulation; and parents can request this information with a quick call to the phone company. Other systems make the service available directly to the parents' own handset, which provides a text read-out of the location.

All parents must dread reading newspaper headlines about abducted children, and these services may help alleviate their fears. But most mobile phone users are unaware that they can be traced by anyone with the authority to do so. In recent notable cases, terrorist suspects have been located in the most obscure hideaways after making mobile phone calls. Helicopter gunship and missile attacks have been made against them in anonymous suburbs, or even en route in their cars, within minutes of their calls being traced. The fun, fashionable phone in our pocket can do more, perhaps, than we realise.

Our other faithful electronic companion is the credit card, as often as not equipped, today, with a tiny silicon chip. Personal details lodged on the chip are supposedly inaccessible to thieves, thereby making the card more secure. On the other hand, the chip also stores the details of electronic transactions, building up a record that may be accessible next time the owner submits the card in order to make a purchase. Streams of information about the owner's purchasing habits are logged and distributed to a wide variety of commercial, marketing and banking databases.

We see news stories about criminals and bomb-makers captured after making rash phone calls. We celebrate the safe child. We become irritated when marketing companies flood our homes with 'personally targeted' junk mail. On the whole, we have taken these unprecedented technologies in our stride, but our relationship with them may change in the future, when we realise more fully the extent to which they use us as much as we use them.

Few of us pretend to know exactly how a mobile phone works, or what all its many electronic components actually do. Voice and text messages are only subcomponents in an array of signals passed between phones, radio masts and a wide range of computers in the network.

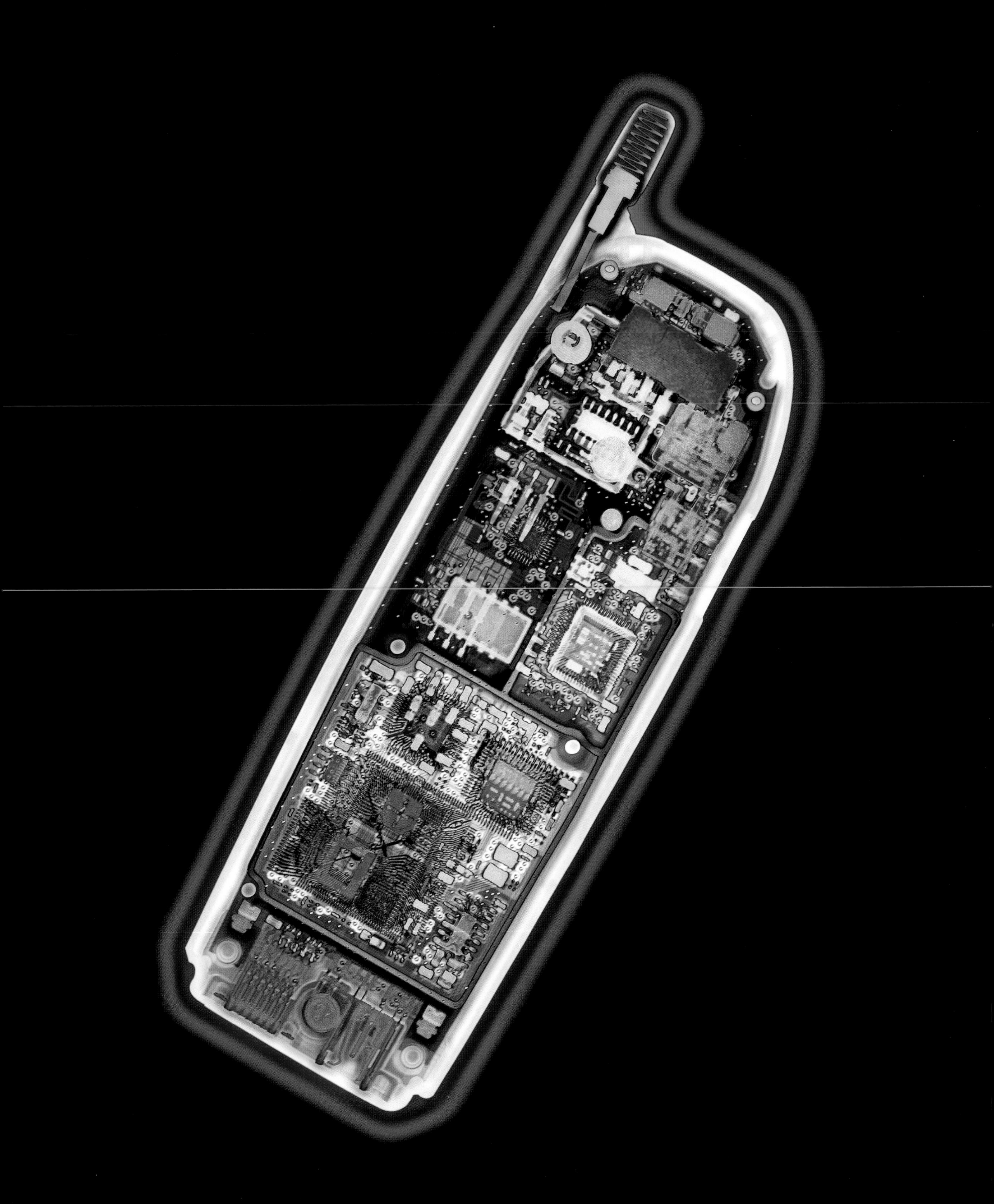

No place to hide

backscatter X-ray technology

Until recently, one of the drawbacks of X-ray photography was that the subject had to be placed between the X-ray source and the detector, as rays needed to pass right through the subject. Now a new approach, called 'backscatter' technology, allows the radiation source and the detector to be on the same side. Passing vehicles, train carriages and cargo containers can be scanned for people, plastic explosives, drugs – anything with an organic chemistry.

Organic compounds tend to have lighter atoms than inorganic ones. Backscatter X-ray machines search for reflected rays coming straight back at them from a target. They distinguish lighter organic materials from the surrounding heavier metals of, say, a truck or a car. Computers then 'tune out' the metal clutter, and highlight just the potentially interesting organic shapes, like humans, or plastic guns, or packets of cocaine. A finely focused beam of X-rays is played across the target, much as the electron beam in a television's cathode tube builds up a picture with closely spaced parallel lines. Each millisecond firing of the beam is answered with a backscatter reflection. Unlike conventional X-ray systems, there is no need for the initial beams to penetrate all the way through the target. This has the added benefit of reducing the dosage of X-rays to levels that are not medically harmful.

This startling image reveals would-be immigrants attempting to enter southern Mexico from Guatemala, hidden in a truck full of bananas. The X-ray scanner was hidden in another innocent-looking truck. It is interesting to see how easily the people hiding inside the illegal vehicle can be spotted. Would we wish to be so easily detectable to authorities hunting for us? It is also comforting to think that bombs, drugs and other dangerous contraband can be more easily detected than has previously been the case. An ever more dangerous world seems to call for increasingly obtrusive interventions.

Illegal immigrants are discovered attempting to enter southern Mexico. The bananas and the hidden people are made clear, because the scanner has been tuned to highlight return signals from organic targets. The walls of the truck are rendered almost invisible.

Living cities

time-compressed images of traffic & people

Major cities are hard for the human imagination to grasp. They sprawl for hundreds of square kilometres, and even their most experienced dwellers cannot hope to know every last corner of them. Our personal map of a city is usually created from snapshot impressions of favourite restaurants and theatres, our places of work, or the houses and streets where our best friends live. We plan our journeys along familiar roads and do not often worry about the countless alternative routes that strangers may have mapped out for themselves. A city is not one thing, so much as a collection of millions of different personalised cities existing in the heads of its inhabitants.

And yet, a city does have an overall shape of its own. Seen in time-lapse photography from a great height – and especially at night – the major transport arteries and population concentrations are revealed. Individual humans in their personalised cars are no longer distinguishable. Instead, we see the secret patterns of a city's life revealing themselves in overall swathes of movement, where the individual car is lost but the vast crowds of cars make their mark, determined not so much by people as by the greater structures and long, accreted history of the city itself.

Urban planners turn to advanced mathematics and 'chaos theory' to try and work out the hidden patterns of a city's flow. A change in just one back road can affect huge areas. Just a few seconds' alteration to the rate at which traffic lights switch from 'stop' to 'go' at a single intersection can make the difference, across the course of a day, between smooth-flowing streets and total log-jam.

We have difficulty imagining the patterns of our lives above and beyond the given moment. We are here, heading to work; we are at our desk; we are, right at this moment, heading back home for supper. The broader significance of our day's movements becomes apparent only when time is compressed, or displayed in a different way than we perceive it; and when our simple, ordinary movements are combined with those of everyone around us, we see bigger, and perhaps less individualistic patterns emerging. Are you one person, or part of a collective society that moves in waves? Is a city inhabited by people, or crowds?

Is the flow of time real, or do we just perceive it that way? These long-duration exposures of Tokyo at night (right) and a busy motorway in Germany (below) show the streak of car lights in busy urban districts. They are portraits of 'traffic flow' through time, rather than of 'cars on a road' at a given instant.

KEI

Big Brother

routine surveillance of our daily lives

George Orwell wrote his dystopian novel *1984* over half a century ago. It has lost little of its power to disturb. Even people who have never read it will be familiar with its key image: Big Brother watching our every move. Televisions beam government propaganda into citizens' homes and, at the same time, the display screens act as cameras. It is all but impossible to avoid surveillance; for Orwell, privacy becomes a vanished concept.

There are now at least 25 million surveillance cameras operating in the world. Britain has the highest figure for any single country: 2.5 million and climbing, at the rate of 300,000 new cameras a year. They watch pedestrians, motorists, railway passengers and high-street shoppers 24 hours a day. The aims of this surveillance are to cut down crime, discourage traffic offences and shoplifting, and generally improve safety for everyone. Security officers can monitor an entire city centre from one control room, using a wall-mounted bank of screens. The irony is that increasing the number of cameras does not necessarily improve the chances of spotting a significant event. Human eyes can tire and the mind's attention can wander, especially when a security guard may be expected to watch two dozen screens simultaneously, hour after hour. Computer software is being developed to help take on some of poor old Big Brother's load automatically. It works by analysing differences in successive images that come up on the screens. Sudden swift movements in a crowd of people might indicate a crime, or a fight; or else an unexpected shape in a railway carriage, which conspicuously fails to move even after the passengers have disembarked, might be a bomb left behind by a terrorist.

Meanwhile, back in our homes, when surveillance does take place, it is most likely to be at our own volition. Landmark legal cases in recent years have involved parents secretly videotaping the wayward behaviour of babysitters, or warring neighbours seeking proof against each other by laying electronic traps. The latest digital video cameras for the home market allow us to become our own Big Brothers and Sisters.

Civil rights organisations are wary of surveillance, but on the whole, the public seems to support it. George Orwell could never have imagined that individuals would turn out to be just as keen on watching each other as the state is on watching them.

A young villain breaks into a car. Is there another explanation for the crowbar in his hand? Is it appropriate to publish or broadcast his image without permission, when he may not have been found guilty in court? In fact, he is an innocent volunteer staging a test of the security system.

14:53:47

28-01-02

5

Echoes in the ether

The interpretation of radar or sonar images requires a different kind of human imagination. It is a leap sideways from the familiar two-dimensional representations of the world that, say, a map or a painting might yield – a kind of vision where scales of distance, rather than the shapes of objects, are the primary sources of information. The latest techniques, however, turn these abstract data into pictures that seem compellingly real.

Finding the range
the classic sweep of a radar

A bar of fuzzy light sweeps its way around a dimly glowing green screen. Where the bar encounters a target, a brief surge of light appears, only to fade slightly once the bar has passed. By the time the bar revisits, the blip has crept noticeably towards the centre of the screen. An enemy missile is closing in …

Radio direction and ranging, better known as radar, has been at the heart of warfare, weather forecasting and the control of aviation traffic for more than half a century. Does this kind of technology produce a 'picture' of the world? In a way, yes. The centre of the screen represents the observer's position, and targets are depicted in terms of distance from that centre, determined by how long it takes for an outgoing radar pulse to be returned. Imagine standing in the middle of a room and turning around slowly, with a hand over one eye to eliminate stereoscopic depth perception. Your surroundings become a 360-degree panorama of shapes, but it is no longer so easy to tell how far away things are. Now imagine closing both eyes and throwing ping-pong balls at chairs, tables and bookshelves, listening for the sound of impact. If a ball bounces off something straight away, that object is nearby, and if the bounce occurs much later, then the ball has gone all the way across the room before hitting something. After a while, you might gain an impression of the room without having to open your eyes.

Our perspective is similar to radar even when our eyes are open. We too are centres receiving information from a 360-degree panorama. Yet the mental model of the world that we build is more complicated than what we see in any given moment, because we seldom stay in the same spot for long. Similarly, when radar sources are allowed to move in relation to their targets (and when computers interpret the data to build a 'model'), something more like a conventional image appears on the screen: an image that is not necessarily similar to how we truly see the world, but very like how we *think* it should look.

As well as listening for the mere fact of the ping-pong bouncing, the well-trained ear might sense other useful qualities, such as the weak, smudgy noise returned by a ball when it hits the cushions of an armchair, the slight double-bounce as it glances off the spines of books on a shelf, or the clean ping of an impact on a wall. In a similar way, radar receivers can sense the differences between what a crisp returning signal should be like, and what actually comes back. It is not just the distances of targets that can be measured, but something of their textures too. Radar, sonar (which uses sound waves) and other echo-location devices have become powerful new ways of seeing.

Preceding pages
A giant dish captures and brings to a focus the almost unimaginably faint traces of radio energy from the deepest realms of the universe. Radio, radar and microwave energies are all forms of electromagnetic light. They can be used to make pictures as well as to transmit information.

Facing page
A radar display in a boat highlights other vessels, quays and rock formations as streaks of yellow and red. The boat's vantage point is represented by the centre of the screen, and all the other information is shown in the form of distance readings from that centre.

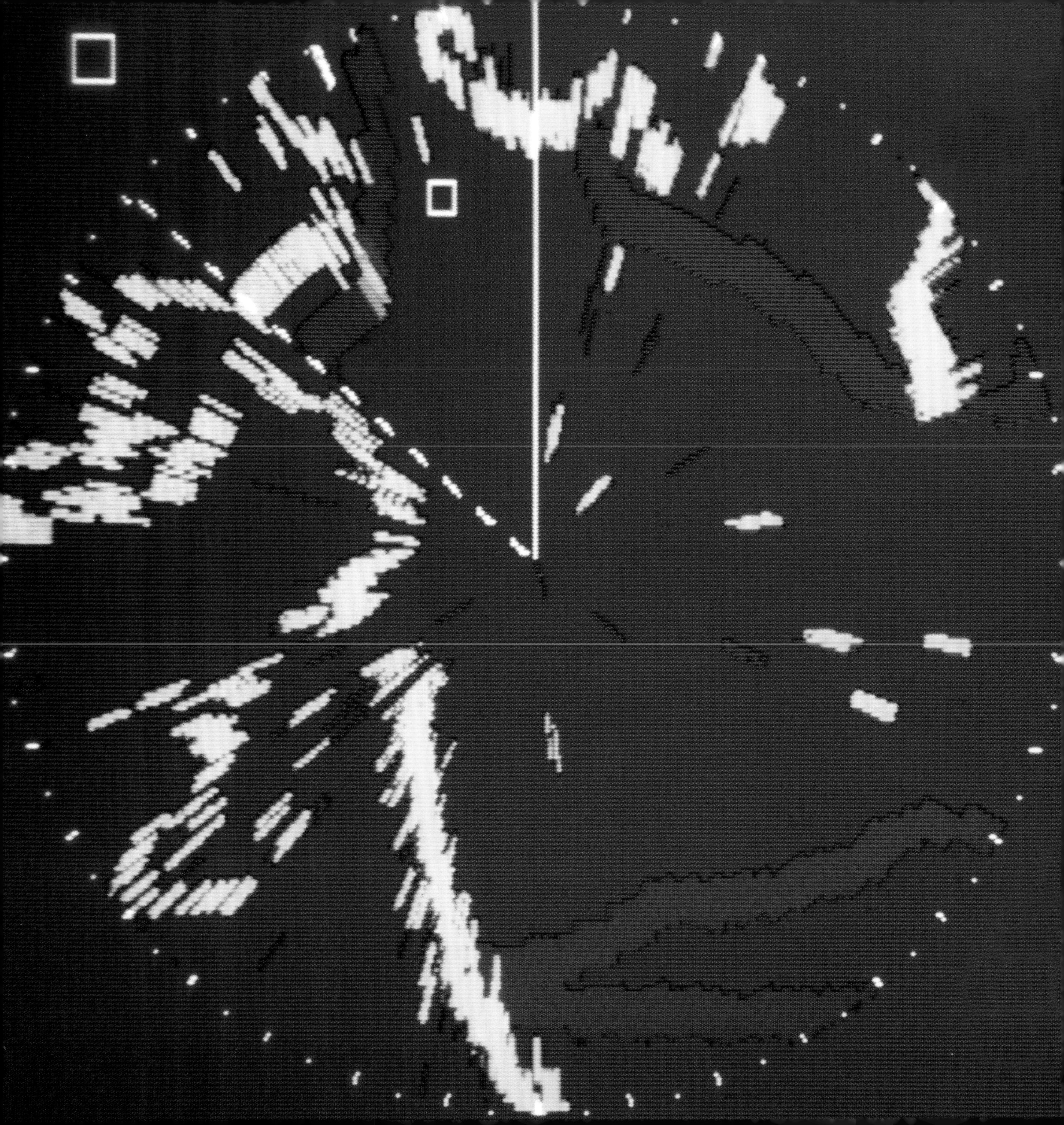

Watching the world

detailed observation from orbit

In 1960, just three years into the age of space flight (the first Sputnik satellite was launched in 1957), an experimental weather monitor called TIROS beamed back crude black-and-white television pictures of cloud tops and storms. Details were hard to distinguish, although the outlines of continents could be discerned, with practice, through the unclouded windows of atmosphere. One day, an image analyst reported that he could just about make out faint patterns in snow, in a picture of a region of northern Canada. At first his report was discounted, but it turned out that lumberjacks had indeed left large tracks in the snow at that very spot, dragging away logs from a forest clearing.

By 1961 the first astronauts and cosmonauts were peering at the Earth through the tiny windows of their capsules. NASA astronaut Gordon Cooper, in his cramped Mercury ship, amazed mission controllers when he reported being able to see roads, buildings and even smoke from chimneys. By 1965 the clarity of the Earth as seen from space could no longer be denied. Astronauts in two-man Gemini capsules corroborated each other's impressions, and returned to Earth with colour photographs of drainage channels, wheat fields – and some interesting airfields and other man-made structures in eastern Europe and Russia. One photo also revealed, with startling clarity, the differences in colour between adjacent agricultural plots in Texas, some of which had received sufficient water to deliver healthy crops, and others where the crops were drier.

These were the kinds of exciting, publicity-generating discoveries made by NASA and the civilian space effort in the first decade of the space age. But behind the scenes, a vast industry – funded just as lavishly as the human space program – was functioning in conditions of such secrecy that even the majority of politicians in Washington were not permitted to know about it. A series of Discoverer satellites was described, in public, as being loaded with biological specimens and science experiments. In fact they contained Central Intelligence Agency (CIA) cameras. Only in recent years has the extent of 1960s spy satellite technology been made clear. Three decades later, much of that technology has filtered into the civilian sector. We now study the Earth from space more precisely than we do on the ground, with infrared, ultraviolet and radar wavelengths, as well as in optical light. We take this multifaceted celestial view of ourselves and our home world more or less for granted today; but the surprise of those early TIROS technicians at the possibilities they had opened up reminds us that such a high vantage point has been available only in recent history.

A beautiful satellite image of the Sahara looks naturalistic, as though it had been captured by a camera in ordinary daylight. In fact the adjusted colours reveal mineral and geological patterns underneath the surface sands, identified by their combined radar and infrared reflections.

Radar archaeology
hunting for the traces of lost civilisations

In 1998, new evidence of a prehistoric civilisation and remnants of ancient temples in Angkor, Cambodia, were discovered by researchers using data from an airborne radar survey conducted by NASA. Angkor is a vast complex of some 1,000 temples covering 300 square kilometres of northern Cambodia. Little is known of the prehistoric occupation of this fertile flood plain, but at its height the city may have supported a million inhabitants. The famous temples unearthed so far (including the huge Angkor Wat complex, visible here in the centre) were built from the eighth to the thirteenth centuries AD and were accompanied by a massive hydrological system of reservoirs and canals. Today, most of the remains of Angkor are hidden beneath a dense forest canopy and are inaccessible due to poor roads, landmines and political instability. This image, about 5 kilometres square, reveals many structures not hitherto depicted on survey maps. The large dark rectangle and prominent walls in the middle of the image are excavated temple walls that have been well-known to archaeologists for many years, but the surrounding tracery of grooves and faint rectangular outlines, especially in the lower middle and right of the image, show where other ancient structures and man-made paths have left faint perturbations in the contours of the terrain. From ground level they are almost impossible to see, because of the dense undergrowth.

The NASA data suggests that Angkor's thick vegetation may conceal many more artificial structures that have not yet been identified. Tuned to penetrate the haze of vegetation and bounce off the hard ground beneath, the radar highlighted a number of circular mounds and undocumented structures. Admittedly, many of these structures are probably less architecturally advanced than the temples already discovered, but they do suggest that estimates for the earliest occupation of Angkor as a major centre of civilisation may need to be backdated by at least 300 years. Even the architecture from the twelfth century and later has yet to be understood in detail: for example, it is possible that the vast network of irrigation channels had a ritual as well as a practical function.

Angkor is situated on the edge of the Tonle Sap lake, a unique body of water that doubles in size during the rainy season. Some ecological or social disaster precipitated Angkor's abandonment around 500 years ago. Perhaps this radar data hides further clues; not only to the mystery of the death of a great Cambodian civilisation, but to the ways in which sudden upsets in the precarious balance between humanity and nature can bring down even the most powerful civilisations.

The huge Angkor temple complex is scanned from space, using radar to penetrate the forest canopy. Faint traces of hitherto undiscovered artificial structures still affect the contours of the ground, despite dense layers of forest that have buried them over the last 500 years.

The ultimate maps

realistic models of landscape via radar

Modern map-making is an essentially fluid process. This three-dimensional view shows the Tigil River on the Kamchatka Peninsula in Russia, in the eastern regions of Siberia. This notoriously cold, inhospitable land is not easy to map from the ground. The raw data for this image was obtained by radar aboard the space shuttle Endeavour, and the final rendition was created by combining radar height and contour readings with optical colour information from separate satellites operated by Italian and German space scientists. Russian geographers also accessed the information. The characteristics of a wild and distant landscape are here revealed as never before.

A traditional flat map is invaluable for navigation, because its scale is precisely in proportion to real distances. We can literally take a ruler and measure on the map how many kilometres' driving our journey will involve. A three-dimensional map, with its arbitrary choice of perspective, may not be so accurate a guide to distances, or even the exact spatial relationships between different places – but it can yield visceral information that flat maps usually lack.

Local authorities in Pasadena, California, for instance, use similar shuttle radar maps to visualise where fires or mudslides pose a risk to buildings. The San Gabriel slopes are regularly ravaged by summer fires, which weaken surface soils by stripping away the grass and shallow roots that tend to bind the soils together. Any serious downpour of rain creates mudfalls, especially in the clefts of narrow canyons. On top of that, Pasadena falls within an area of southern California prone to earthquakes. These potential dangers are easier to grasp from a three-dimensional map: the looming mountains both shelter and threaten the city below, and it is the overall character of the landscape that is conveyed, rather than merely its peaks and troughs at any given coordinate.

Map-making has traditionally been a painstaking craft. Now radar topology (automatic analysis of contours and textures) allows detailed maps to be generated in hours by computers, to be presented in a variety of ways: flat plan, oblique view, or profile of the terrain.

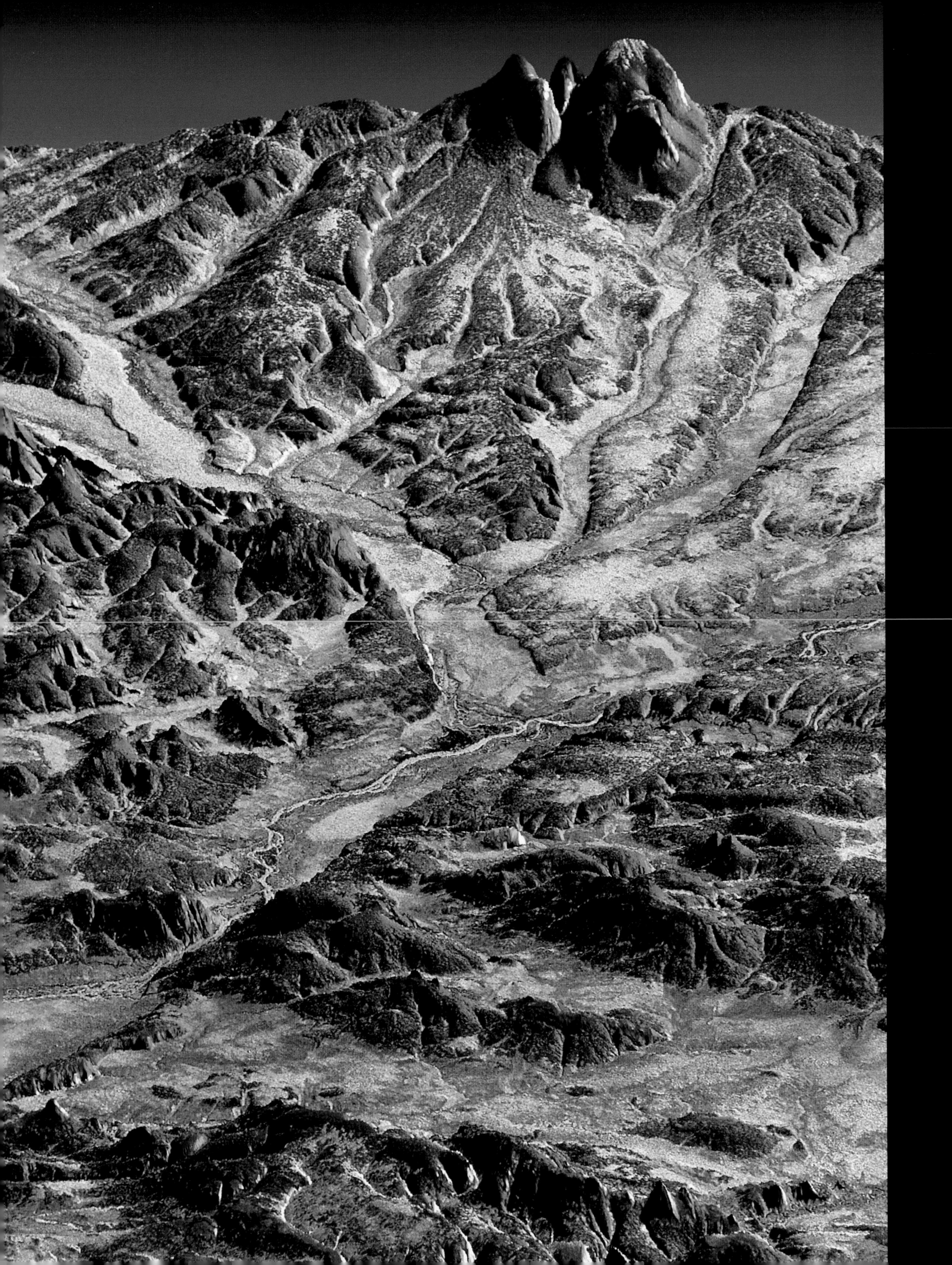

Crowded skies

air traffic control versions of the outside world

London's Heathrow Airport began its career in 1946 as a simple airstrip despatching 9,000 flights a year to fewer than 20 destinations. Now it handles 1,300 planes per day, feeding 160 destinations around the globe. A plane takes off or lands once every 45 seconds, and the giant terminals process 64 million people a year. Meanwhile, the skies of North America swarm with 50,000 planes every day, passing through 21 separate air traffic control zones.

An air traffic controller needs to be aware of dozens of planes simultaneously, with each plane denoted on a radar screen by a code sign. That code, in turn, refers to a 'progress strip', sometimes electronically displayed, but often literally consisting of a strip of paper bearing relevant information that cannot easily fit on screen: the type of plane, its radio call sign, its destination, the assigned route and altitude, and so on. The progress strips for a departing plane are passed from ground controllers who handle taxiing and runway queues to air controllers monitoring take-off, and then on to other specialists handling the plane's transition from local to international air space. The global complexity of handling air traffic is staggering. There are myriad control centres, all running different radar and computer equipment – some of it dangerously incompatible and outdated. Industrial relations between employers and overworked controllers are seldom healthy, and the entire process is always threatening to fall apart.

Traffic control mistakes are surprisingly rare, but terrible when they do occur. In 1996 a Saudi Arabian jumbo jet collided with a Kazakh-owned cargo plane near New Delhi, killing 349 people in the world's worst mid-air accident. The ground can be just as hazardous. Two jumbos collided on the same runway at Tenerife Airport in March 1977. One of the planes missed a crucial taxiing exit from the runway in thick fog, and the other plane slammed into it at full take-off speed. Nearly six hundred people lost their lives. In 2001, two Japan Airlines planes came within a hair's breadth of the worst mid-air collision in aviation history. A DC-10 and a Boeing 747 jumbo jet approached within ten metres of each other high above central Japan.

It is a wonder that so few collisions have occurred so far – but this could change. The busiest airports now want to ease congestion by shrinking the time lag between take-offs to 30 seconds, and there are moves afoot to allow planes at cruising altitudes to fly over or under each other's paths with a safety margin of just 300 metres. This picture of an air traffic controller's screen conveys some of the awesome responsibility of the job.

This snapshot of an air traffic controller's radar screen shows dozens of individual aircraft as yellow blips, each with a code number. Radio 'transponders' in all large commercial aircraft transmit a unique identifying signal, but many small private planes are not yet required to do so.

RDK013P
ROSUN
UAL953
UPTON
GOLES
UKA12A
DENBY
AUR678
HAYDO
BAW35Y
AVB2SM
RRR1547
LOVEL
DAYNE
EZS941
REXAM
NOKIN
EGCC_STAFA
BMA245
UAL931
ROBIN
BRY38B
MONTY
STAFA
SKINA
MYNDA
TELBA
NITON
BAW797B
BRT429
SAPCO
EGCC_TRENT
BAW27B
CRX867P
EZY047
BAW1655
ELGAR
GROVE
ABHON
SSW501P
WELIN
TC_WELIN
RADNO
MADLI
BAW1746
RYR336
WOBUN
WLC202
BRT347
SHT7Z
EZY355
BUZAD
RYR2835
TALGA
BAL117A
CHELT
DIKAS
BCN
CUMRI
GRSCJ
BRT112
HEN
BNN
NUMPO
RYR804
WOTAN
SIREN
BASET
BUNCE
BRY9CD
CPT
WOD
BAW766
VIR23
EIN168
TAMEF
EXMOR
TUTON
BAW742
GODAL
OCK
VAPID
PEPIS
ADSON
MID
PEPIS_CORRIDOR
KLM742
BRY5GW
BAW4503
BRY06A
BRIPO
GIBSO
BAW2466
BEWLI
BAW1793
AVANT
POMPI
FPG087
BOGNA
NEDUL
ELDER
KATHY
THRED
TINAN
DAWLY
CAMRA
WAFFU
BEECH
HARDY
SFD
HOLLY
WILLO
TC_WILLO
TIMBA
EZS973
LARCK
UKA1RZ
MAY
TIGER
DET
KLM789
BRT618
BAW2939
RYR983
BUSTA
RYR107
STOAT
LOREL
ADMIS
AFR1160
BMA6XC
SUDBY
BRAIN
MATCH
JAL401
WESUL
FERIT
GILDA
BCY5024
BAW855
SAY787
BAW818
BMA817
BAW34F
HEMEL
DRAGN41
DRAGN42
MOGLI
GDE513
N66SG
LESTA
UAL941
CTR
XP

Lighting up the night
humanity's spread across the planet

Most of the images in this book are derived from non-optical techniques, but the pictures on pages 6–7 and opposite are fascinating exceptions. They show the distribution of lights visible on Earth at night. A space shuttle orbits the world once every 90 minutes, and the planet itself turns once every 24 hours. In order to obtain a complete night-time map, successive shots of the night-shaded side of the Earth, gathered during a multi-day shuttle mission, are montaged into a single image. The camera that makes the scans is not a simple photographic machine. It is a precision electronic instrument that accurately calibrates the different intensities of light. Hence, the final montage distinguishes between the particularly brilliant concentrations of lights in major cities, as against the sprawling but slightly less dense suburbs, or the tiny spark-like glimmers of remote rural towns and villages.

Just a century ago, much of North America was frontier land. Now the towns and suburbs have spread into the heartland, and even into the driest deserts. Canada is still largely wilderness, while western Europe is almost entirely occupied. China and northern Russia have vast territories that are not suited to large-scale human occupation, while the Indian subcontinent betrays its teeming population. South America is well-lit in its coastal regions. Africa shows just how far removed it still is from the world of wealth – and how much of its interior is hostile desert; and in Australia, few people have any interest in occupying the dry, hot interior, and all the major cities huddle around the coast.

We can read this image of night-time western Europe in various ways: as a map of wealth distribution, a welcome show of life and civilisation in an otherwise vast and lonely universe, or as an environmental warning sign. In one sense, it must be delightful to see the glittering lights through the windows of a space shuttle. On the other hand, the tell-tale signs of human encroachment across such vast reaches of a small and fragile world might give even the most homesick astronaut pause for thought.

Western Europe depicted by the lights of its urban communities at night. Brightness levels in the final image accurately convey the relative light output of the various towns and cities. Burn-off flares from oil rigs in the North Sea are also clearly visible.

Seeing the wind

visualising dynamic weather systems

Since the dawn of the space age, satellite instrument makers have prided themselves on their ability to see through atmospheric haze, thus providing an unhindered view of the Earth's surface. Investigating the atmosphere itself is a more subtle problem. Suitably tuned radar systems can distinguish heavy cloud formations because of the high water droplet concentrations, but they are not so good at analysing the motions of clear air. This QuikSCAT satellite image showing wind intensities and directions is derived from radar data – not of the air, but the sea.

A radar 'scatterometer' working at microwave frequencies analyses the reflectivity of small wind-caused ripples (called cats' paws) on the surface of the water. The radar pulses are fired obliquely, rather than straight down: they are on the hunt for the sides of the ripples, not their crests. An ocean surface returns a stronger signal in a strong wind, because the ripples are higher, and their sides reflect more of the radar energy back towards the scatterometer antenna. A smooth ocean surface returns a weaker signal because less of the energy is reflected. In addition, the scatterometer distinguishes the different polarisations of the outgoing and returning pulses: the vertical or horizontal orientation of the electromagnetic waves. The resulting view of wind vectors and intensities appears marvellously self-confident, but the scatterometer can only tell us about what is happening to the air within an area of approximately ten metres above the ocean, where the wind has a direct effect on the surface. Winds at higher altitudes, moving in completely different directions, may not leave such easily identifiable traces on the water.

In this particular wind map, the North American continent is clearly visible as a dark silhouette at the top right; South America is just visible below it. These shapes represent absences of oceanic wind data rather than 'images' of the continents. Storm systems are gathering just south of Alaska, and in the southernmost reaches of the Pacific Ocean.

Wind speeds and directions over the Pacific Ocean mapped by a Seasat satellite in September 1989, with wind speeds defined by different colours. The blue regions are experiencing gentle breezes, while red and orange lines show wind speeds of 40–70 kilometres per hour.

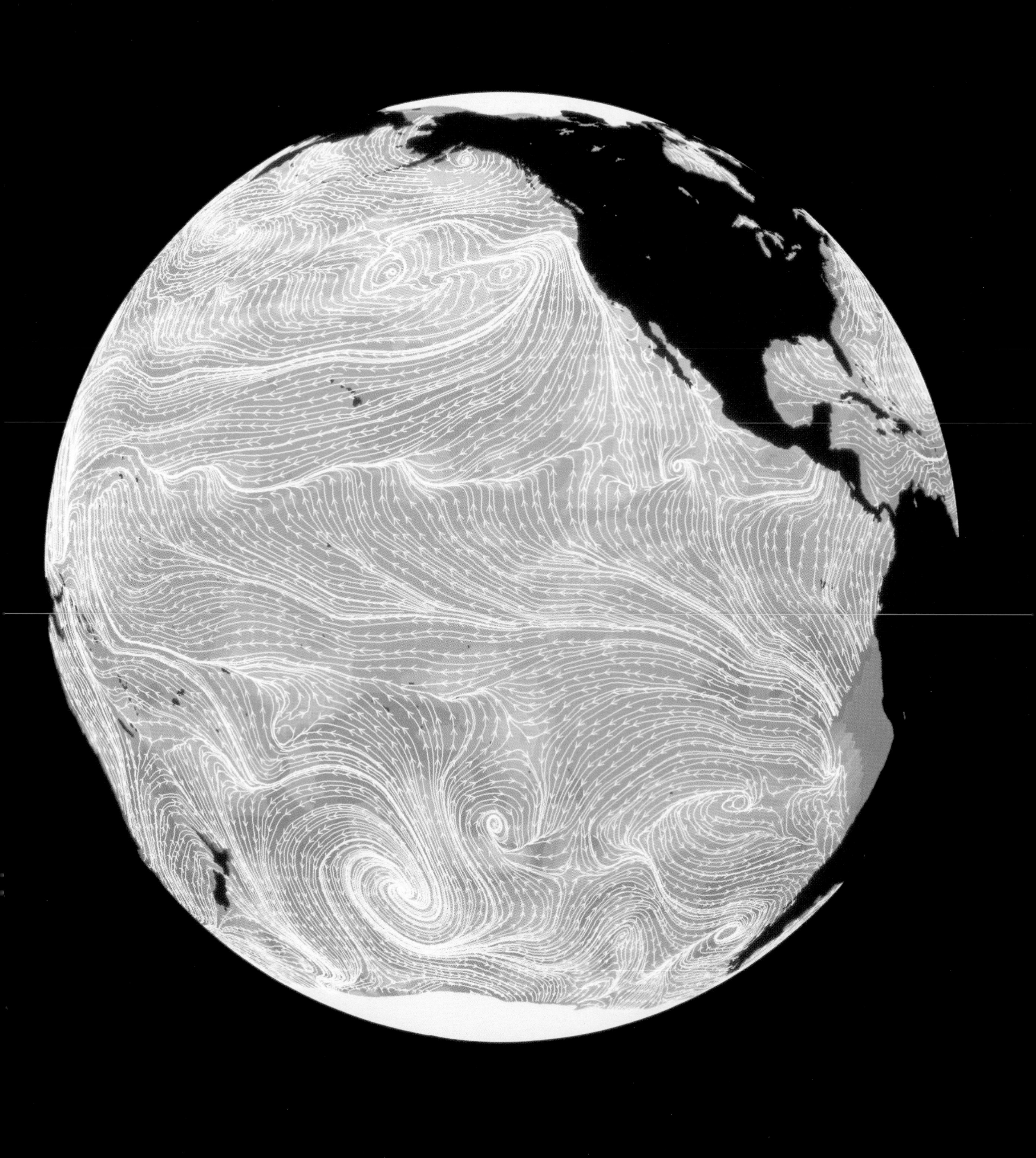

A storm brewing

images of extreme weather

These views of Hurricane Isidore were taken by a NASA satellite on 20 September 2002. After flooding huge areas of western Cuba, Isidore was upgraded from a tropical storm to a Category 3 hurricane. Sweeping westward to Mexico's Yucatan Peninsula, it caused major destruction and left hundreds of thousands of people homeless. Although weakened after passing over the Yucatan land mass, Isidore regained strength as it moved northward over the Gulf of Mexico.

The left-hand image seen here is a visualisation of cloud extent, in which infrared data is combined with optical imagery. The white clouds of the main storm formation are unambiguous; the blue areas are those deemed statistically likely to be clouded; and red shows areas that are definitely clear of cloud, where the warmth reflected from underlying land masses is clearly visible to the satellite's instruments. The middle picture combines images taken at slightly delayed intervals, allowing a stereoscopic impression of cloud heights to be computed. The right-hand image isolates 'albedo', a precise measurement of sunlight reflected from the cloud tops. All these factors play a role in creating and dissipating hurricanes, but the mechanisms are not yet fully understood.

We cannot predict hurricanes as far ahead of time as we might like. Every year between early June and late November, hurricanes threaten the eastern and gulf coasts of the United States, Mexico, Central America and the Caribbean. In other parts of the world, the same types of storm are called typhoons or cyclones. These huge storms wreak havoc when they make landfall. They can kill thousands of people and cause billions of dollars' worth of property damage. Typically, they start as modest thunderstorms; and there is no telling which ones might develop, after many hours or even days, into full-blown hurricanes. Warm, humid air from the ocean surface begins to rise rapidly. The water vapour condenses to form storm clouds and droplets of rain; the condensation process releases heat into the upper air; and a tremendous convection pressure is set up, with warm air from sea level rushing to occupy the cooler air pockets above. The cycle continues, eventually creating a gigantic storm, where winds swirl at more than 100 kilometres per hour around a small central 'eye' of calm conditions and clear skies.

Simultaneous displays of a violent weather system, viewed in different wavelengths, allow meteorologists to study the interactions of sunlight, air humidity and underlying geography in the development of a storm.

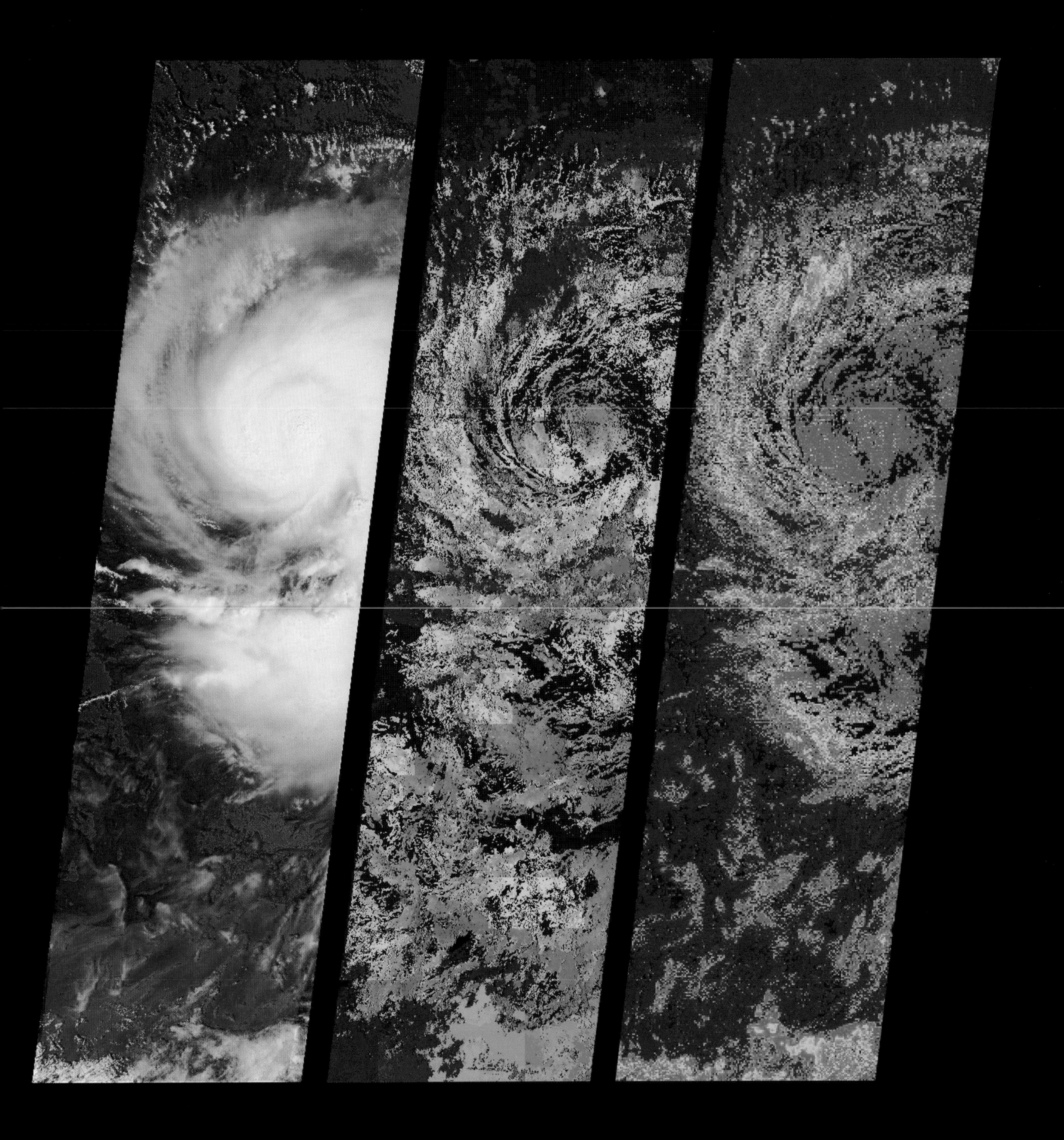

Warning signs

chemical clues reveal climate change

Free oxygen normally exists as two oxygen atoms bound together. For every million oxygen molecules in our atmosphere, there will be perhaps a dozen molecules where three oxygen atoms have combined, rather than the normal two atoms of conventionally stable oxygen. This chemical rarity, known as ozone, is created by the disruptive effects of ultraviolet sunlight; yet these fragile traces protect life on Earth from that very same light.

Most of the ozone resides in a layer of atmosphere 10–40 kilometres above the ground (the stratosphere). Every year, during the southern hemisphere's September–November spring season (while the northern hemisphere is in its autumn months), atmospheric ozone over the Antarctic continent is rapidly destroyed by chemical processes. Over the course of three months, approximately half of the ozone disappears; at some altitudes, the losses approach 90 per cent. Once the peak three-month period of depletion has passed, ozone layers recover – but the extent of that recovery may not now be keeping pace with the destruction.

This alarming seasonal trend was first recorded in 1985 by Joseph Farman, Brian Gardiner and Jonathan Shanklin of the British Antarctic Survey. Now widely known as the 'ozone hole', it has become a major international concern. Over the last two decades, the average size of the hole, in terms of ground area, has expanded to 10 million square kilometres, encroaching the tip of South America, sweeping over the Falkland Islands, and thinning the ozone levels over New Zealand. The annual depletion period sometimes stretches from August to December before regeneration kicks in. The net result is a gradual rise in harmful ultraviolet radiation levels reaching the ground. There have been marked increases in human cataracts and skin cancers during our generation, almost certainly attributable to ultraviolet sunburn. Valuable plant crops and forests may similarly be under threat, along with many amphibian and marine species whose body tissues are particularly sensitive to radiation damage.

Ozone destruction seems to be caused by human activity on the ground, and by the passage of countless thousands of aircraft through the stratosphere. Chlorofluorocarbons (CFCs) are the worst offenders. These chemicals, which consist of atoms of chlorine, carbon and fluorine, are used in aerosol sprays, refrigeration systems and countless other industrial processes. They leak out of the machinery, or are carelessly released when old hardware is scrapped. Slowly, they drift upwards into the stratosphere. When ultraviolet radiation hits a CFC molecule, chlorine breaks off and attacks and destroys ozone.

This is certainly one 'invisible' phenomenon that we need to be able to depict clearly. The NASA image of the ozone hole on the opposite page, made during 1998, is derived from radar and spectrometer readings, with chemical variations colour-coded, and ozone levels (or rather, the lack of them) rendered in three dimensions for ease of interpretation.

Ozone concentrations over Antarctica are depicted on a colour-coded scale, from red (the highest levels) through yellow and green, to blue (the lowest levels). The smaller image shows warm water currents threatening the ice sheets around the Antarctic coast.

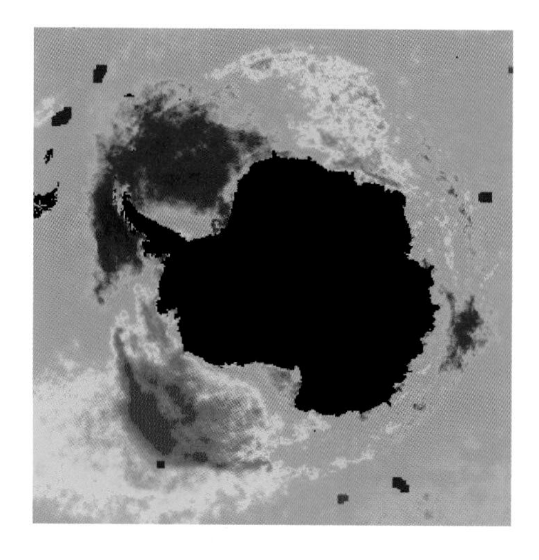

Into the abyss

mapping the ocean floors with sound

The renowned futurologist Arthur C. Clarke wrote, 'How inappropriate to call this planet Earth when quite clearly it is an ocean world.' Two-thirds of the planet is covered in water. The average depth is 3.5 kilometres, but in some areas the seabed is three times further down into the abyss. Underwater mountain ranges dwarf the Himalayas. Deep ocean thermal vents support bacteria which thrive in boiling water. Countless unknown animals prowl the darkness. We have mapped most of the surface of Mars to a resolution of a few metres, but the oceans of our own world are still largely unexplored territory.

Sound waves are the principal tool for investigating the ocean floor. Short-wave ultrasound is very good for resolving fine detail and is often used for charting shallow reaches, but after a certain depth it is absorbed by the water. Only longer-wave pulses have any chance of reaching the deeper seabed. However, they are not so good for mapping in fine detail. Small areas can be mapped fairly well by ships dedicating several days to the task, but the greater ocean expanses have been resolved only very approximately to date.

Space scientists have found a way to contribute to underwater mapping. Variations in average sea surface height, measured by radar, reveal where the Earth's gravity field is particularly large or small. Gravitational anomalies are usually associated with sea floor ridges and troughs, and differing densities of crustal material. At the Scripps Institute in California, researchers are cross-referring the latest satellite radar information with a haphazard variety of depth-sounder surveys made over the last 30 years. Their aim is to produce the best overall map of the sea floor yet attempted. Meanwhile, conventional sonar techniques are improving – becoming 'multi-spectral', rather in the style of electromagnetic sensors. Oblique backscatter technology (where signals are scattered sideways from a target) gives clues about surface textures and materials, while the slight differences between signals that are bounced cleanly from the seabed, and those that are partially absorbed by soft sediments, are also taken into account.

Ocean waves constantly interfere with sonar surveys, causing a survey ship to pitch, roll and yaw, throwing off individual soundings by several metres. Motion sensors can now record these lurches 100 times per second with an accuracy of 0.1 per cent, and feed their compensating data into computers. In addition, very accurate measurements of sea water properties are gathered throughout the survey so that allowances can be made for sound refraction caused by water-density variations. Depth-sounding *sounds* as though it should be fairly straightforward – but it is not.

This sonar scan shows the East Pacific Rise, a fast-spreading upswell in the Earth's crust formed by the collision of two tectonic plates. Below, another region of the Pacific floor, where gulleys plunge thousands of metres into the bedrock, and grand underwater mountains protrude.

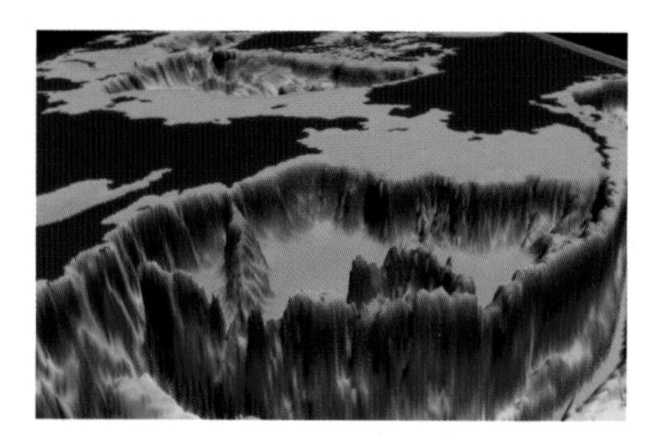

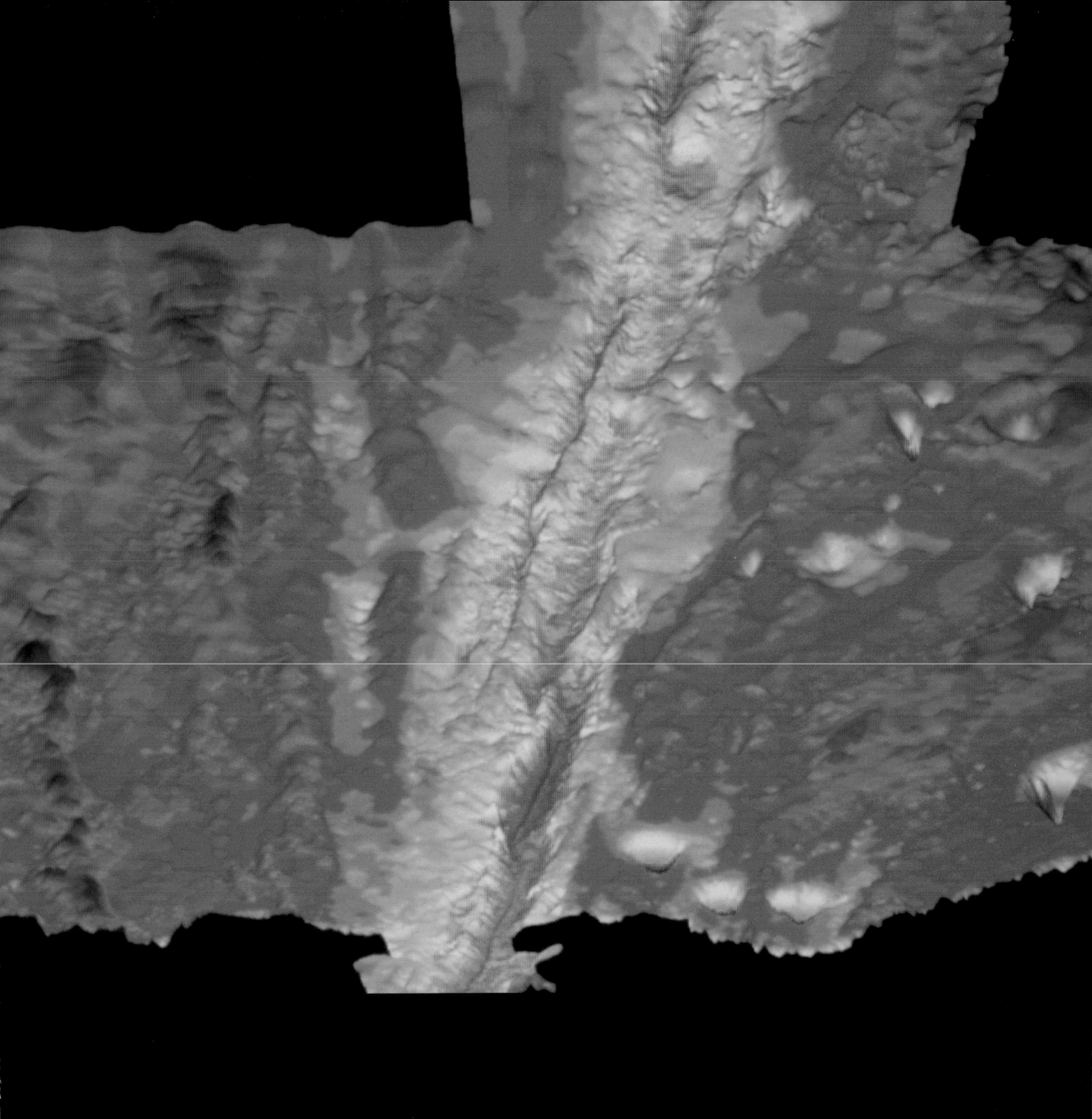

Hidden impact

seismic modelling of impact craters on Earth

Phil Allen, a geophysicist based in Scotland, discovered a hitherto unknown meteorite impact crater on Earth by pure chance. In the summer of 2002 the petroleum company BP asked him to look at three-dimensional computer enhancements of seismic data from a gas field four kilometres below the North Sea. During his analysis, Allen discovered unusual features in layers of chalk lying above the gas field. A deformation in the chalk looked like a crater, but Allen did not quite believe it until he met with Simon Stewart, a BP geologist who confirmed his findings. The crater, named Silverpit in honour of a nearby sea floor trench, is 140 kilometres off the east coast of Britain. An asteroid 200 metres across crash-landed 60–65 million years ago, gouging a pit three kilometres across. The collision would not have been sufficiently violent to scatter debris across the planet, but it would have created a huge tidal wave.

Seismic studies of the Earth are difficult to conduct, because only the lowest-frequency sounds can penetrate the crust to any distance. Low-frequency pulses produce low-resolution data. Typically, a survey requires many soundings from different locations and directions to be cross-related, and few organisations can afford surveys of any detail. Oil and gas companies are notable exceptions, often spending millions of dollars on commercially promising sites. BP needed to understand the rocks beneath the North Sea at the Silverpit location, and it just happened that a significant crater was discovered there.

Allen and Stewart were surprised that the ancient impact had been preserved in such fine detail. A kilometre-thick coating of fine sediment overlaying the chalk has protected what might otherwise have been a very fragile crater. Intriguingly, it dates from around the same time that a much larger meteorite struck the Yucatan Peninsula in Mexico, possibly bringing the age of dinosaurs to an end. Was the Silverpit meteorite a fragmentary accomplice in that greater catastrophe?

The Silverpit story highlights how much we have yet to discover about our home world, despite a feeling among some commentators that we have discovered almost every last nook and cranny of the planet. The way in which the crater was revealed also tells us that we only know about those phenomena that we happen to come across. What other surprises lie in store in the Earth's geological record that we have not thought to look for, or yet been able to find?

The Silverpit crater is not particularly grand, but its obvious bowl shape has survived because that particular region of the North Sea has remained relatively undisturbed over the last 60 million years. Traces of the Yucatan impact are harder to identify because too much has happened to the Earth's land and sea across that huge area in the last 60 million years, and the crater walls have long since been obscured by erosion. But the deepest layers of crust were compacted by the original impact, as though by a hammer, and variations in rock density affect its gravity field to this day. A colossal asteroid impact best explains the circular contours of the density variations detected by a satellite survey.

Traces of a gigantic crater shown by a gravity scan of the Yucatan Peninsula. The crater rim is seen as the larger green, yellow and red ring while the white line shows today's coastline. The image below is from a seismic survey of the Silverpit crater recently discovered in the North Sea (data, facilities and software supplied by Perenco, PGL and Landmark).

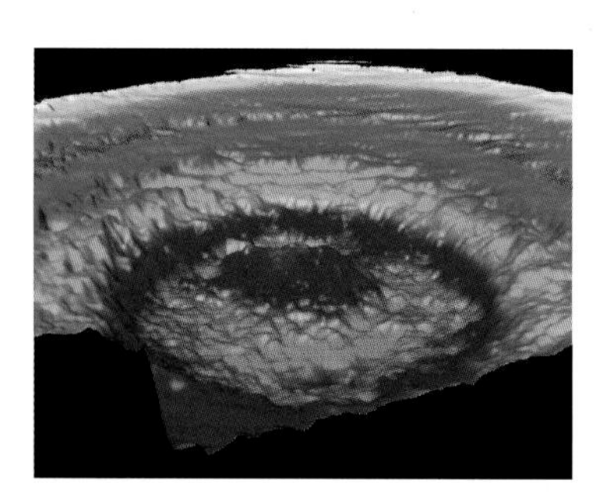

Cracks in the fabric

movement stresses in the Earth's crust

We know that the Earth undergoes tremendous upheavals, yet they occur so slowly, over such immense stretches of time, it is almost impossible for us to perceive them during our relatively short lives. Almost but not quite. Those who have escaped an exploding volcano or survived an earthquake soon abandon their illusions about the solid permanence of the ground they stand on. However, for most of us – and most scientists, too – the chances of witnessing dramatic events in geology are few and far between. Our planet's epic violence is usually too subtle to encroach on our daily sense of reality. The latest space-borne instruments reveal what we cannot see from ground level: the extremely subtle traces of geological change over time.

Satellite 'interferograms' are derived from pairs of overlapping radar images, taken weeks, months or even years apart. Computers identify slight differences of altitude, or minor lateral displacements, by analysing variations in the data between one scan and the next. The notorious Hayward and San Andreas crustal faults in California are closely monitored for any signs that new earthquakes might be imminent. Most shifts in the crust are caused by harmless variations in groundwater levels, but Californians cannot be too careful. San Francisco and Los Angeles will never be free of the significant possibility of catastrophe.

Earthquakes are notoriously difficult to predict, and geological upheavals can happen with terrifying unexpectedness. This image shows an interferogram derived from radar scans taken just five days apart. A European Space Agency remote sensing satellite passed over the Hector Mine in California on 1 October 1999. The next day, there was an earthquake centred on the site, measuring 7.1 on the Richter scale. A second scan, taken five days later, highlighted newly created contours in the surrounding crust. Each band of colour shows ten-centimetre displacements left by the quake.

Shifts in the Earth's crust are detected after an earthquake. Two satellite radar scans of the terrain, taken five days apart from precisely the same orbital coordinates, fail to match up when superimposed on the ground. Mismatches in the data show stresses in the crust.

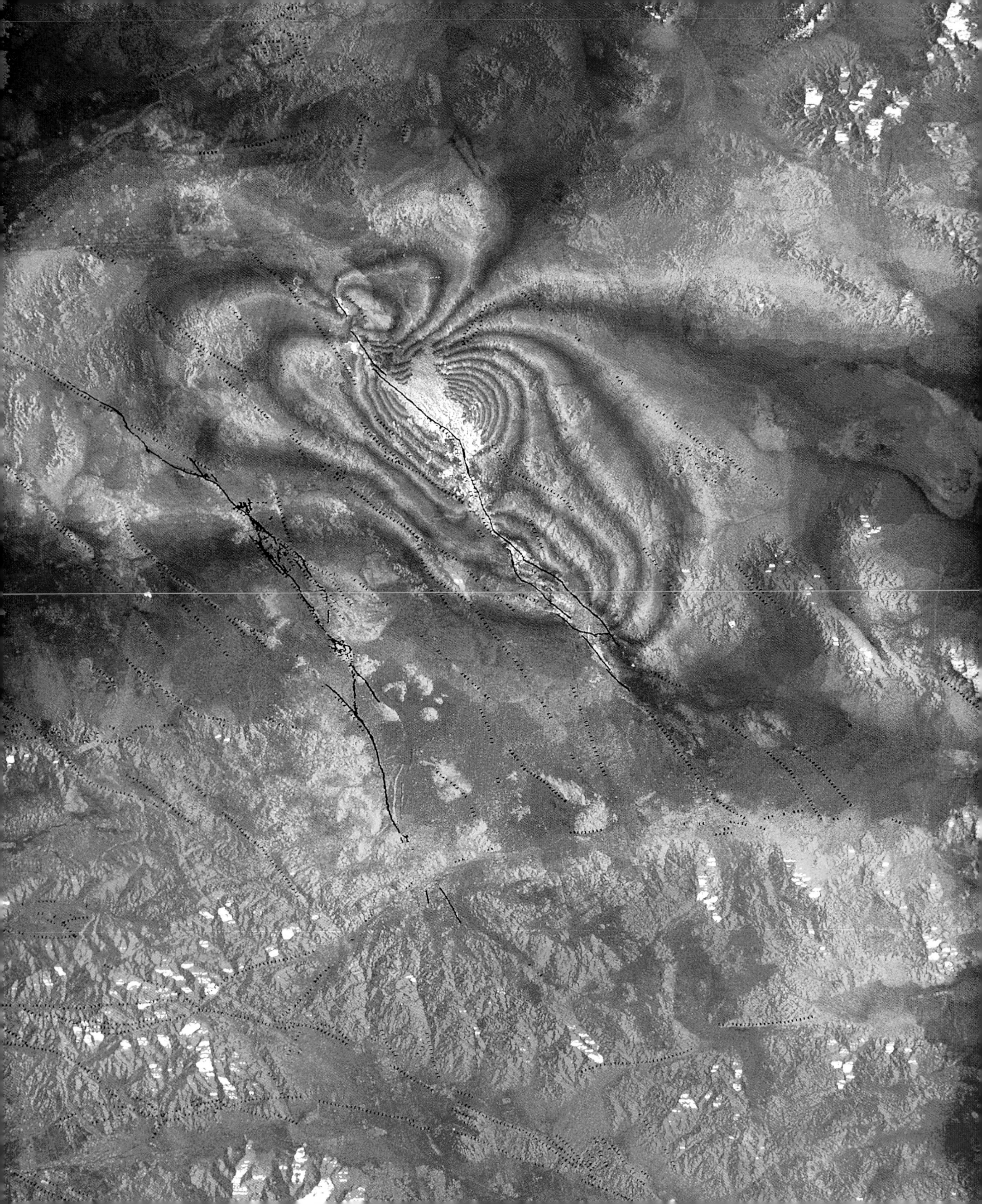

Prospecting for gold

space spectroscopy of mineral deposits

When the African and Eurasian tectonic plates ran into each other about 80 million years ago, the collision slowly destroyed an ocean. Its sedimentary bed was folded, crumpled and pushed upwards to form the Atlas and Anti-Atlas Mountains of Morocco.

This image was acquired on 13 June 2001 by the Advanced Spaceborne Thermal Emission and Reflection Radiometer (ASTER) on NASA's Terra satellite, with additional hardware contributed by Japanese and European scientists. The instruments work with 14 spectral bands, from the visible to the thermal infrared wavelength regions, and can resolve features as small as 15 metres across. One set of instruments probes the borderline between visible light and near infrared, another looks for the thermal energy in sunlight reflected from the Earth's surface, and a third makes the subtle distinction between different types of infrared – that reflected from sunlight, or that emitted by the terrain's inner warmth. This 'multi-spectral' approach allows observers to distinguish between different minerals, because each shows a characteristically different ratio of reflected and emitted energies.

When this information is colour-coded, a beautiful map of hidden chemistry emerges, and the Atlas Mountains reveal their secret lives. The yellowish, orange and green areas are sedimentary limestones, sandstones, clays and gypsum from the ancient seabed, and the dark blue and green areas are granite rocks from the deeper planetary crust.

Similar techniques are used today to identify diseased crops. Although a healthy leaf might look, to our eyes, pretty much the same as a leaf which is subtly diseased but not yet brown or curled around the edges, the infrared signatures of leaves are markedly different when seen from space as a collective population. Given sufficient warning, farmers can sometimes take action to save healthy crops before a blight has taken hold. Many organisations on the ground, both commercial and academic, can download and analyse satellite images such as this at relatively low cost, and debate their meanings. The best balance, for instance, between efficient exploitation of natural resources and our desire to protect the environment, is not necessarily made clear by any given image, no matter how precise the pictures from space might seem. Final processed images are often defined on the ground, by the (perfectly legitimate) highlighting of one type of information over another.

The Atlas and Anti-Atlas Mountains of Morocco are scanned for their underlying minerals. Yellow, orange and green areas show soft sedimentary clays and sandstones from an ancient seabed that formed the mountains. Dark blue and green show hard granite rocks from the deeper crust.

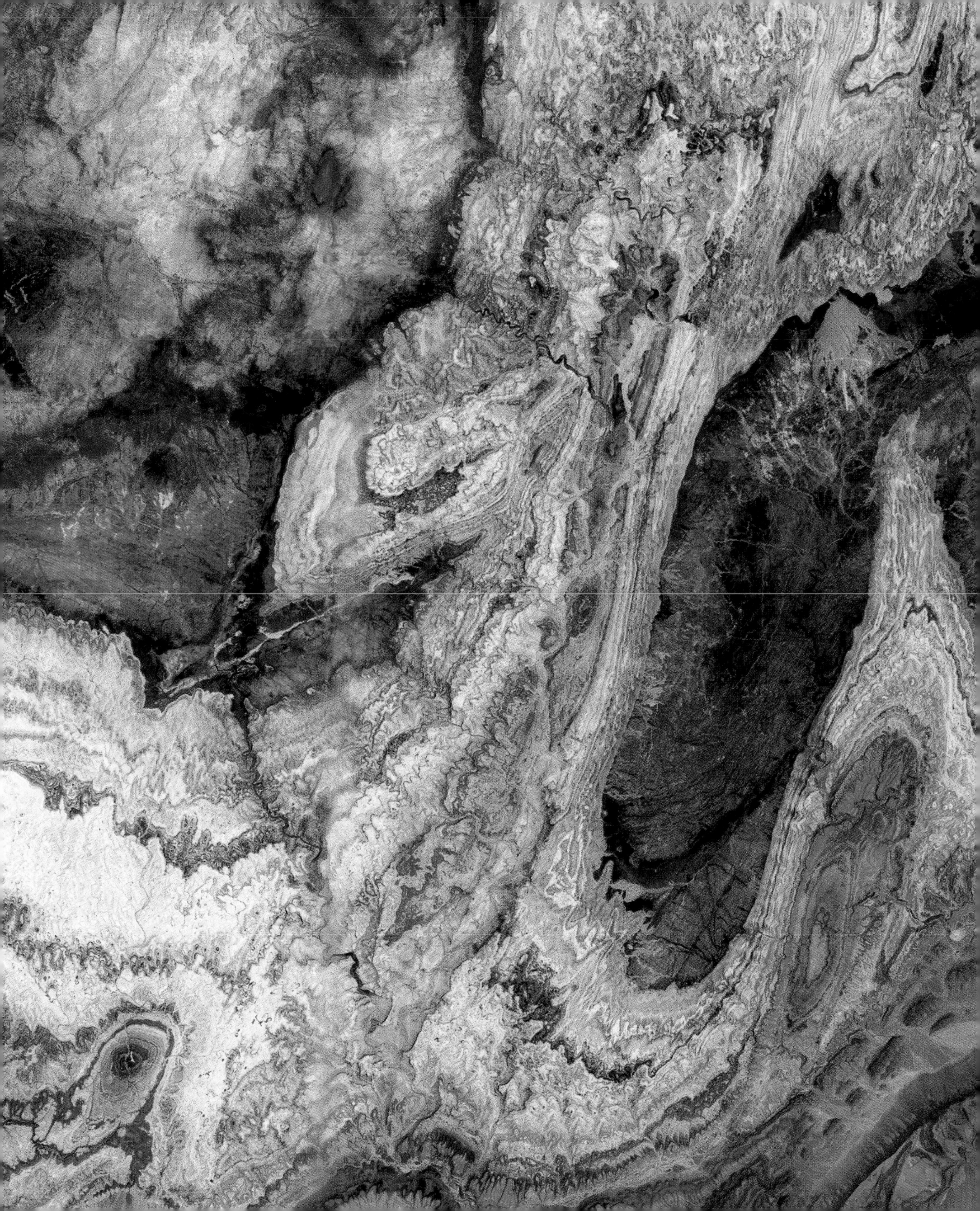

Bumps in the field

unevennesses in the Earth's gravity

The Earth's gravity is lumpy. Rock is denser than water, and water in its liquid form is denser than ice. Large concentrations of densely compressed rock, such as in a mountain range, have greater mass than low-lying coastal sedimentary rocks, and so on. Since the Earth has such varied geological features, the planet's mass is not evenly distributed around the globe, and consequently, neither is its gravitational field.

Gravity is usually studied using instruments that determine how hard the ground pushes back against a precisely known weight. It is a slow process. Mapping from space would be better. The orbits of all artificial satellites are affected by unevennesses in the Earth's gravity, but those effects are too small to detect easily. Comparisons have to be made between the paths and altitudes of many different spacecraft, recorded over a number of years, before useful patterns can be discerned. This data is usually extrapolated from other experiments that have little directly to do with gravity research: ground radar tracking of spacecraft to monitor their orbits, for instance, or the laser altimetry of a craft whose original purpose was to map the ground.

In March 2002 a US/German mission was launched to speed up the process and hunt specifically for gravity variations. The Gravity Recovery and Climate Experiment (GRACE) uses two identical satellites, flying exactly the same orbit, and separated from each other by a distance of exactly 220 kilometres. As the leading satellite encounters regions of slightly stronger gravity, it is pulled away from its trailing companion. Fluctuations in gravity are measured by monitoring the distance between the two satellites to an accuracy of one micron (about one-fiftieth of the width of a human hair). GRACE has already obtained accurate maps of the Earth's gravity field, but it is not enough to make just one set of maps, for the pattern of the field is ceaselessly variable. Spring melting of ice fields, redistribution of surface and ground water, shifting of the Earth's molten core, and tidal influence of the moon all change the field day by day. GRACE will produce new maps every 60 days, over a five-year period, to help scientists understand the overall dynamics of the gravity field over time.

A computer analysis of the Earth's gravity field, centred on the Atlantic Ocean. The colours show differences between the theoretical value of the planet's gravity compared with actual measurements. Green is the 'normal' value, while red shows above-average field strengths.

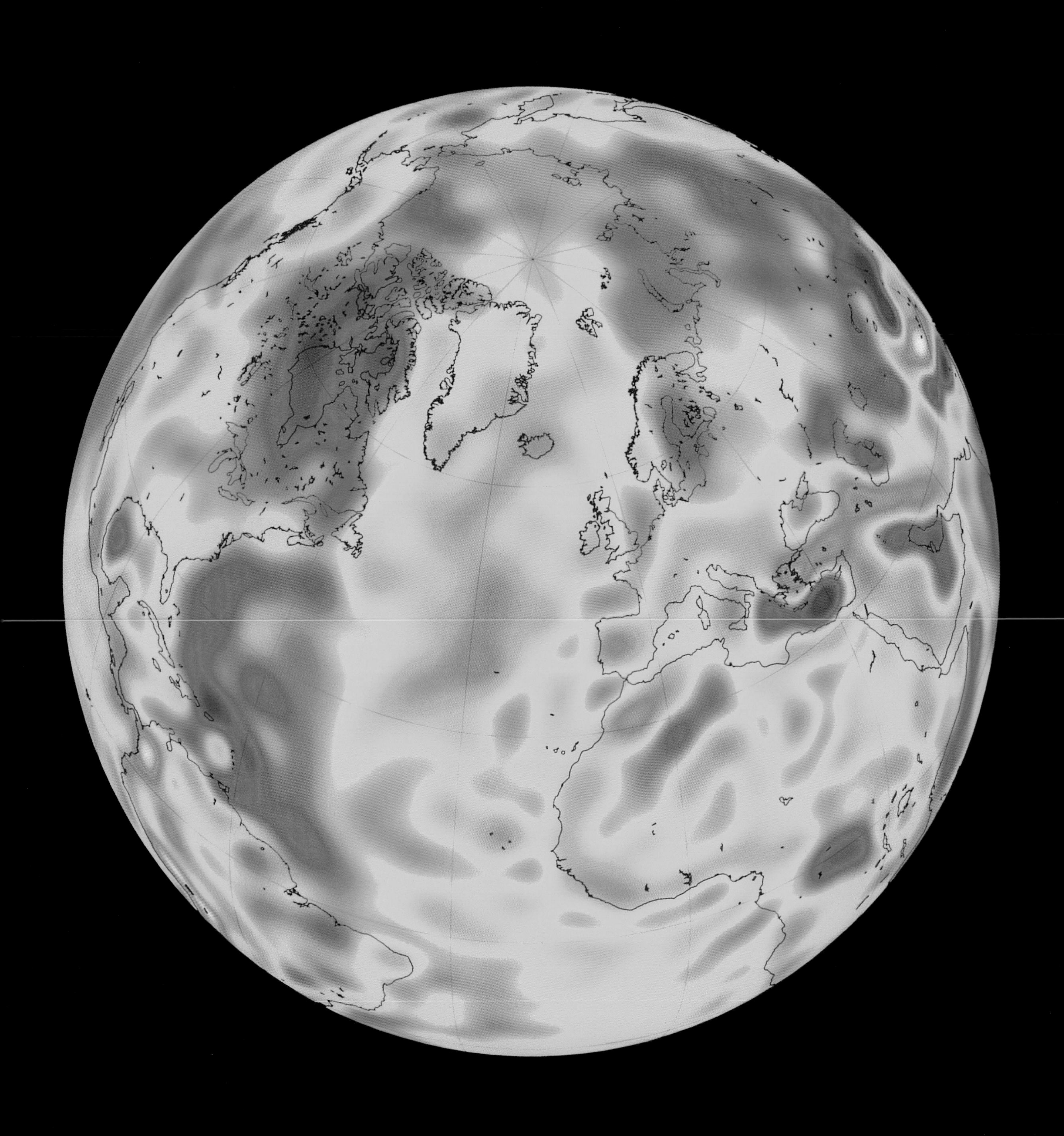

6

Our place in the cosmos

We have probed the hitherto unknowable structures of the Sun with X-rays and ultraviolet, and tested the surface of Mars with exotic spectrometers. We have identified the death screams of black holes and mapped gamma ray explosions that are almost certainly the most powerful surges of energy in the universe. A satellite working at the microwave region has mapped the earliest phases of creation: the 'echo' of the Big Bang itself. As more data comes in, and yet more detailed images of the sky are created, so the mysteries of cosmic evolution become ever more profound.

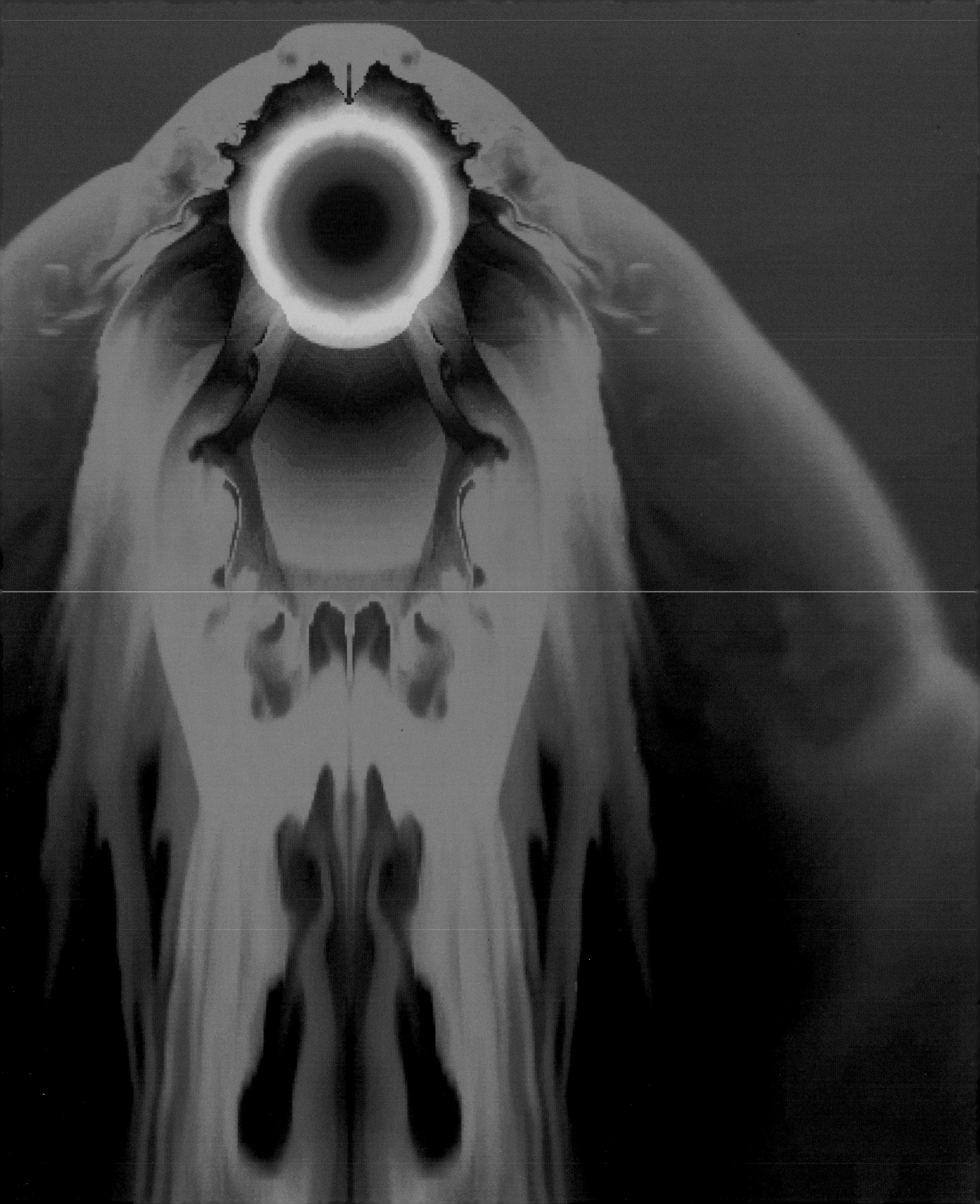

The delicate balance

the Earth as a living entity

In the late 1960s, NASA scientist James Lovelock was part of a team investigating the possibilities for life on Mars. As he pondered the differences between the inert Martian atmosphere and the much more dynamic atmosphere of our own planet, he realised that a fine balance exists on Earth as a result of constant feedback between living and non-living systems. He initially suggested that life maintained the oxygen-rich composition of the air with unusual precision, and few biologists would argue with that. But by 1979 he had broadened his ideas to include the climate, the rocks and the oceans in a cohesive self-regulating system. He argued that the Earth's overall biomass, though consisting of millions of different species, exerts a collective environmental influence that supports the survival of life. Whether or not this is 'purposive' is a matter for philosophers, but our modern understanding of the Earth does indeed support the idea that it is an integrated system rather than just a ball of rock with some life crawling around on it. The planet does seem capable of regulating itself. At least, it has done so until recently; the balance may have been disrupted by excess carbon dioxide generated from human activity.

Some scientists reject the idea that the Earth might be alive, even in a metaphorical sense. Gaia is a romantic name for a complex dynamic system that reacts to itself, and is self-stabilising to within remarkably fine tolerances. We are used to discovering chemical and temperature regulation in biological organisms. We have also encountered plants, animals and bacteria that exist in complex multi-species alliances to their mutual survival benefit. We call them 'symbiotic' organisms. Earth may be a super-symbiotic system on a grand scale. Words like 'life' and 'organism' are loose terms, the definitions of which are still hotly debated. All we can say is that the Earth demonstrates some characteristics that would not surprise us if we observed them in a living creature. Perhaps a hitherto invisible pattern governing the planet's overall behaviour is at last becoming apparent to us.

Preceding pages
A complex computer simulation demonstrates the massive shock wave around one of the two stars in a binary system, shortly after its companion star (above the image, out of the frame) has exploded in a 'supernova'.

Facing page
A composite representation of the Earth, showing surface temperatures and cloud cover, with cloud heights exaggerated for ease of reference. This is one snapshot from a model that depicts, over time, broad dynamic patterns in the environment.

Looking back

can life on Earth be detected at a distance?

We are extremely keen to discover whether or not other planets harbour life, but there is a major problem to be solved before we can answer this question. Space scientists cannot reliably detect life on the one world where its signs should be glaringly obvious: the Earth. If they cannot find it here, they will have a tough job finding it anywhere else.

When the Galileo space probe swooped by the Earth in 1990, building up gravitational momentum for its long trip to Jupiter, all its instruments were pointed towards us for a unique experiment. Strong absorption of light at the red end of the visible spectrum, particularly over the continents, indicated the presence of chlorophyll, the molecule essential to (mainly green) plant life and photosynthesis. Spectral analysis of sunlight passing through the Earth's atmosphere revealed a high oxygen content. Since oxygen is extremely reactive, a dead planet could not support free oxygen in its atmosphere for very long. It takes life (plants especially) to replenish it constantly. Galileo also spotted small quantities of methane in the atmosphere. As far as we know, methane can only be the product of biology.

However, in August 2003 a perplexed NASA team reported their failure to sense life in the Atacama Desert of northern Chile. Admittedly, it is one of the world's driest deserts, but it is teeming with life nonetheless. Scientists on the ground were pestered by flies, even as they marvelled at the variety of lichens growing on and under rocks, or watched vultures circling over their heads – a sure indication that plenty of other animals had to be around somewhere. But colleagues at the Ames Research Center in California, poring over photos and instrument data transmitted from the field scientists, never detected anything they considered an unambiguous sign of life. These results were unsettling for scientists trying to develop robotic landers and instruments for planetary exploration. If an animal walked in front of a Mars rover, few observers would be left in doubt as to the presence of life – but what if it turns out to be sparse, harder to detect? What subtle data signatures, beamed to us on fragile radio links from an unimaginably remote alien planet, might prove the existence of life?

The Galileo space probe looks back at Earth in 1990, on its way to Jupiter, with the Moon in the foreground. In an age of advanced astronomy and space exploration, we still cannot be sure if life is prevalent throughout the universe, or if it is unique to this solar system.

Alien landscapes

Venus mapped by radar

Venus is permanently clouded by a dense carbon dioxide atmosphere, and no surface features are ever visible through optical telescopes or space cameras. The atmosphere traps heat, and a runaway 'greenhouse' effect has made this a hellish world where ground temperatures are high enough to melt lead. The atmosphere and weather on Venus are reasonably well understood, but the only clues we have about the planetary crust are derived from radar.

This perspective view of Venus, generated in 1997 by computer from Magellan space probe data, shows part of the lowland plains in Sedna Planitia. Altitude measurements reveal the contours, but the computers have also been programmed to colour-code the different intensities and directions of radar reflection. This gives some clues about the geology of the planet, as well as the lie of the land. Circular depressions with associated fracture patterns, called 'coronae', are apparently unique to the lowlands of Venus, and tend to occur in clusters along the planet's major tectonic belts. According to NASA scientists, the bright flow lines in the foreground are evidence of extruded volcanic magma flows, but the big central depression is a mystery. It may or may not lie below the surrounding plains, and may or may not be surrounded by a raised rim, or a moat outside the rim. Perhaps an upsurge of magma pushed the crust into a blister, before falling back and allowing the blister to sink into this depression? This image makes Venus seem even more of an alien and mysterious world.

Until very recently, radar data from space took some weeks, or even months, to analyse and translate into different kinds of picture, depending on what scientists wanted to highlight. Raw data from long-defunct probes that flew in the 1970s and 1980s is often analysed afresh by modern computer software on Earth, to render it in ways that were not possible in past decades. Now, relatively inexpensive space probes with off-the-shelf electronic components can beam back what appear to be state-of-the-art mineral analyses and lifelike stereoscopic images, generated and distributed around the world on the the same day of transmission. No doubt a future generation of scientists will extract yet more information from the raw data, by applying software of as yet unimaginable sophistication.

The Magellan space probe mapped most of the Venusian surface by radar in the late 1990s. The planet is permanently swathed in thick cloud from pole to pole, and very little information about its surface can be obtained with ordinary optical telescopes and cameras.

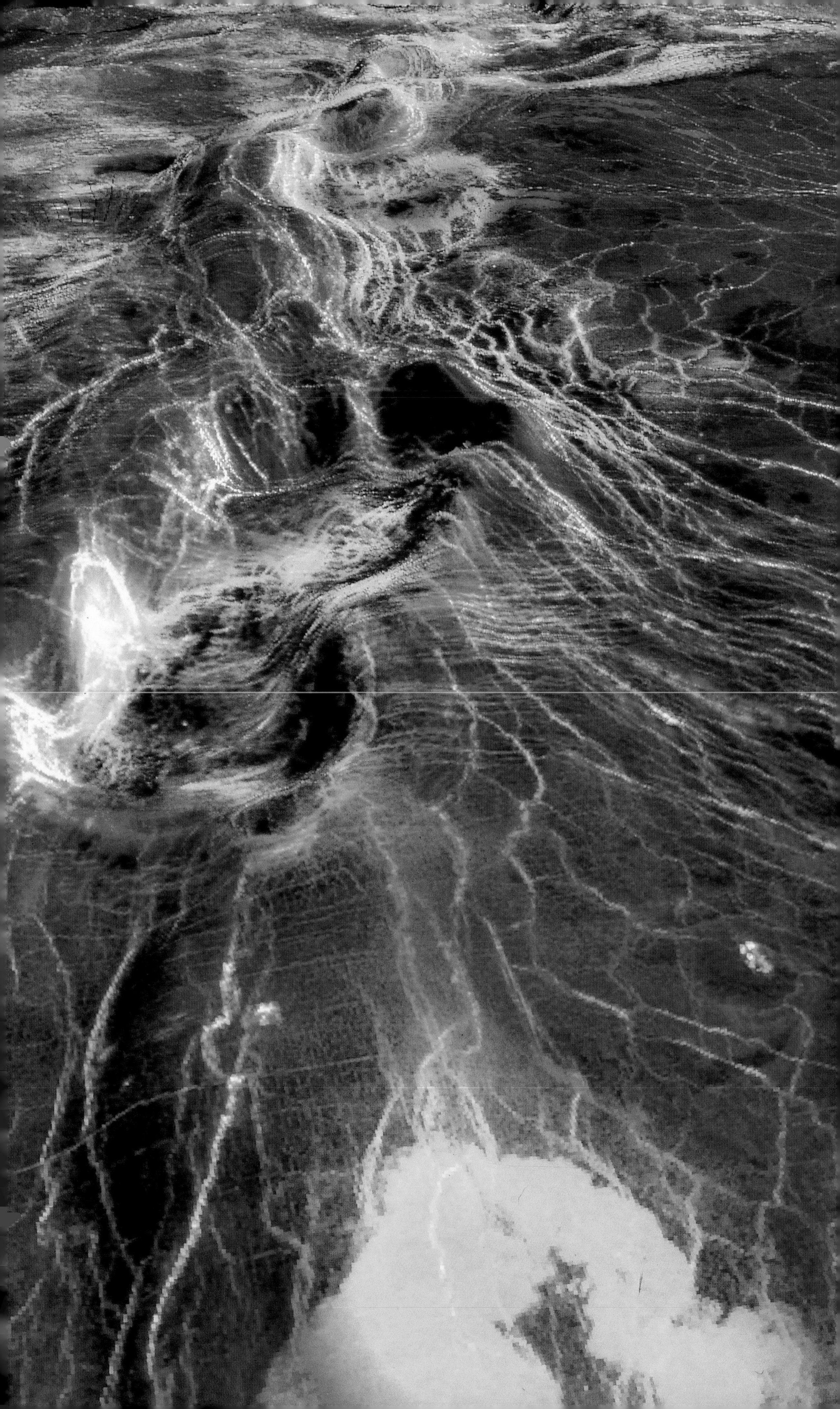

As good as being there

virtual models of Martian terrain

The first Mariner-class spacecraft to fly past Mars in the 1960s returned grainy monochrome pictures of a planet far less interesting than scientists had hoped. There were craters – too many of them – suggesting a world that lacked dynamic erosion forces. Mars seemed just another dead rock, similar to our Moon, only bigger. However, subsequent Mariners found the tips of colossal volcanoes, deep ravines and channels that looked vaguely like dried-up rivers. Mars was not a dull place after all. In the mid-1970s, two Viking landing craft and their orbiting mother ships beamed back colour images of great clarity.

How was NASA to keep taxpayers on side, and willing to pay for such missions, as public interest gradually faded? Until the 1990s, the business of exploring Mars, or space in general, seemed largely a matter for scientists. Now the internet has transformed the public's relationship with space. In 1996, the Mars Pathfinder landing probe, and its little wheeled rover Sojourner, were headline news around the world. Much of the attention focused on the virtual, interactive nature of the mission, and NASA websites received a record-breaking number of hits that has not been surpassed to this day. The controllers at NASA viewed the planet virtually, via twin cameras on the spacecraft whose lenses were set apart by the same distance as a pair of human eyes, and all images and data transmitted from the Martian surface were made accessible to anyone with a home computer. The Red Planet was no longer far away in the depths of space; it was right here on Earth.

The Global Surveyor and Mars Odyssey orbiting craft have transformed our view of the planet by sending thousands of detailed pictures. Surface features the size of a small car can be resolved, and beautiful and unusual geological features have been discovered, along with fascinating hints of sedimentary layers in river-like channels, and the suggestion that water might be trickling down the sides of some gulleys even today. Meanwhile, non-optical spectrometers are scanning the mineral composition of Mars, conjuring up as many mysteries as answers.

It is relatively simple, today, to create virtual models of Martian terrain, because such a quantity of raw data has accumulated over the last 30 years. Mars now seems a familiar place, and we can all explore it through our telephone sockets.

This computer model shows a section of Valles Marineris, a network of canyons that wraps its way around Mars for thousands of kilometres. The main trenches are 200 kilometres wide, and their steep walls plunge 7 kilometres below the crust of the surrounding plains.

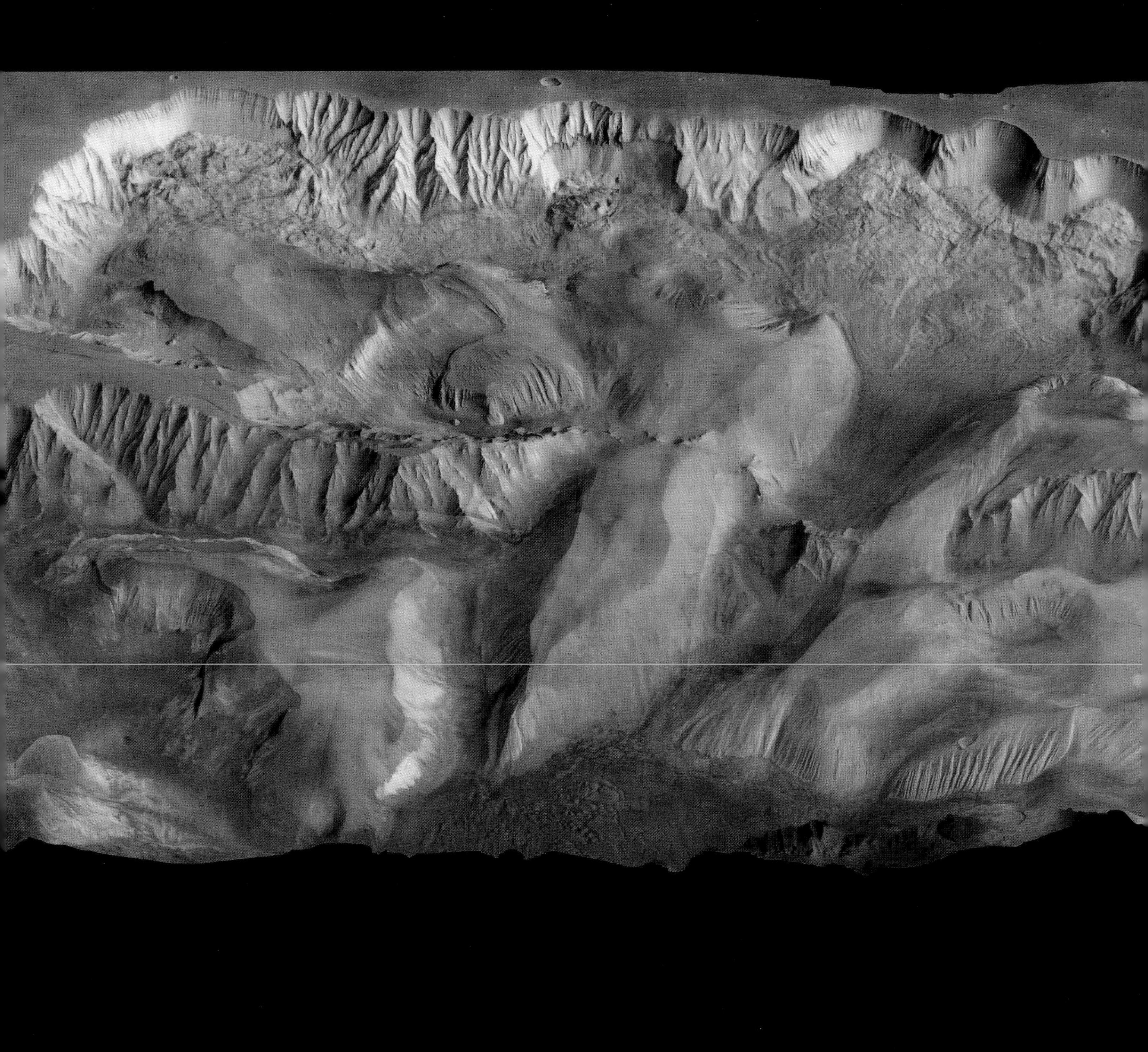

Water on Mars

intensive chemical mapping from orbit

The shortest wavelengths in the electromagnetic spectrum belong to gamma waves, the highly energetic products of decaying radioactive atoms, or high-speed subatomic collisions. The universe is awash with cosmic rays – highly energetic particles hurled out from a variety of sources. When these rays encounter Mars, they are unimpeded by its thin atmosphere, and they smash directly into atoms in the soil. Neutrons are released, which scatter and collide with other atoms, which in turn emit gamma rays in order to shed the extra energy they have absorbed. In 2001, Mars Odyssey's gamma ray spectrometer distinguished about 20 different elements in the soil, including hydrogen.

We have been able to analyse the distribution of elements in the first metre depth of Mars's topsoil, but only individual atom types, and not necessarily all the molecular compounds, which have to be inferred from a variety of clues. In 1975, a pair of Viking landers touched down on Mars and conducted some ingenious analyses of surface soil, looking for signs of carbon-based life. The soil was highly reactive when placed into contact with organic compounds carried aboard the landers, but after many weeks of argument, scientists gave up on trying to find surface life and ascribed the reactivity to a volatile inorganic chemistry in the soil – perhaps something along the lines of hydrogen peroxide. A quarter of a century later, we are still some way from a definitive understanding of Martian soil.

The picture on pages 2–3 is one of the most emotionally and politically charged images to have emerged from planetary research. It implies that essential drinking water might be bountiful on Mars, and in such quantities that we could also realistically consider extracting its oxygen for breathing and its hydrogen to fuel our spaceships. So this is not just a clinical image of possible water distribution on or near the planet's surface, delivered in a scientifically dispassionate manner; it is also the expression of a hope that we might one day be able to inhabit Mars. The image opposite is a computer simulation of what could happen, one day, if we were able to tinker very slightly with the Martian atmosphere and make the average temperature rise to just above the freezing point of water. Surface ice could be liberated, and liquid water might pool in the Martian craters.

Climate models are combined with studies of primitive algae that might be able to survive on Mars. In theory they could transform the thin atmosphere over a period of several centuries. Planetary reserves of ice would turn into liquid water as temperatures rise.

Restless giants
the electromagnetic lives of planets

Imagine the electrical generator in a car that charges up the battery: it consists of a coil of copper wire rotating inside a ring of magnets. That movement between an electrical conductor and a magnetic field causes electrical energy to surge through the wires. This is the process behind a dynamo. A similar mechanism operates inside planets, generating a powerful yet completely invisible force around them. As a planet turns on its axis, its molten, iron-rich interior rotates at a slightly different rate to the outer mantle of solid rock, and a huge dynamo effect comes into play, creating doughnut-shaped magnetic fields around the planet. These, in turn, trap atomic particles, either from space or from the planet's upper atmosphere.

In 1958 James Van Allen discovered intense radiation looping around the Earth in wide belts: a direct result of high-energy particles trapped in the magnetic field. Named in his honour, the belts were a cause for concern in the early days of space exploration: the radiation is medically hazardous to humans in their capsules, and can even destroy electronic circuits inside robot probes. Fortunately, most spacecraft tend to orbit beneath the 1,000-kilometre altitude where the innermost belt's radioactive influence becomes a hazard.

Jupiter and Saturn are mantled in gas rather than rock, but the variation in rotation rates of their outer and inner layers produces magnetic fields just the same, with colossal radiation belts to match. The grandeur of these planets as seen through optical telescopes is insignificant in comparison to the hidden power of their surrounding electromagnetic energies.

By contrast, the magnetic field around Mars has been shown to be very weak. From this unexpected negative result, planetary scientists have concluded that there is little internal drag. Mars's interior may still be quite hot, but the magma is more sluggish today than it was 4 billion years ago, when both Earth and Mars were young. The Red Planet's invisible aura is fading.

A supercomputer models Earth's magnetic trail, its passage through space and its deformation by the 'solar wind' of charged particles thrown out by the Sun. The interactions are vast and complex; the energies can be lethal; yet this phenomenon is invisible to the naked eye.

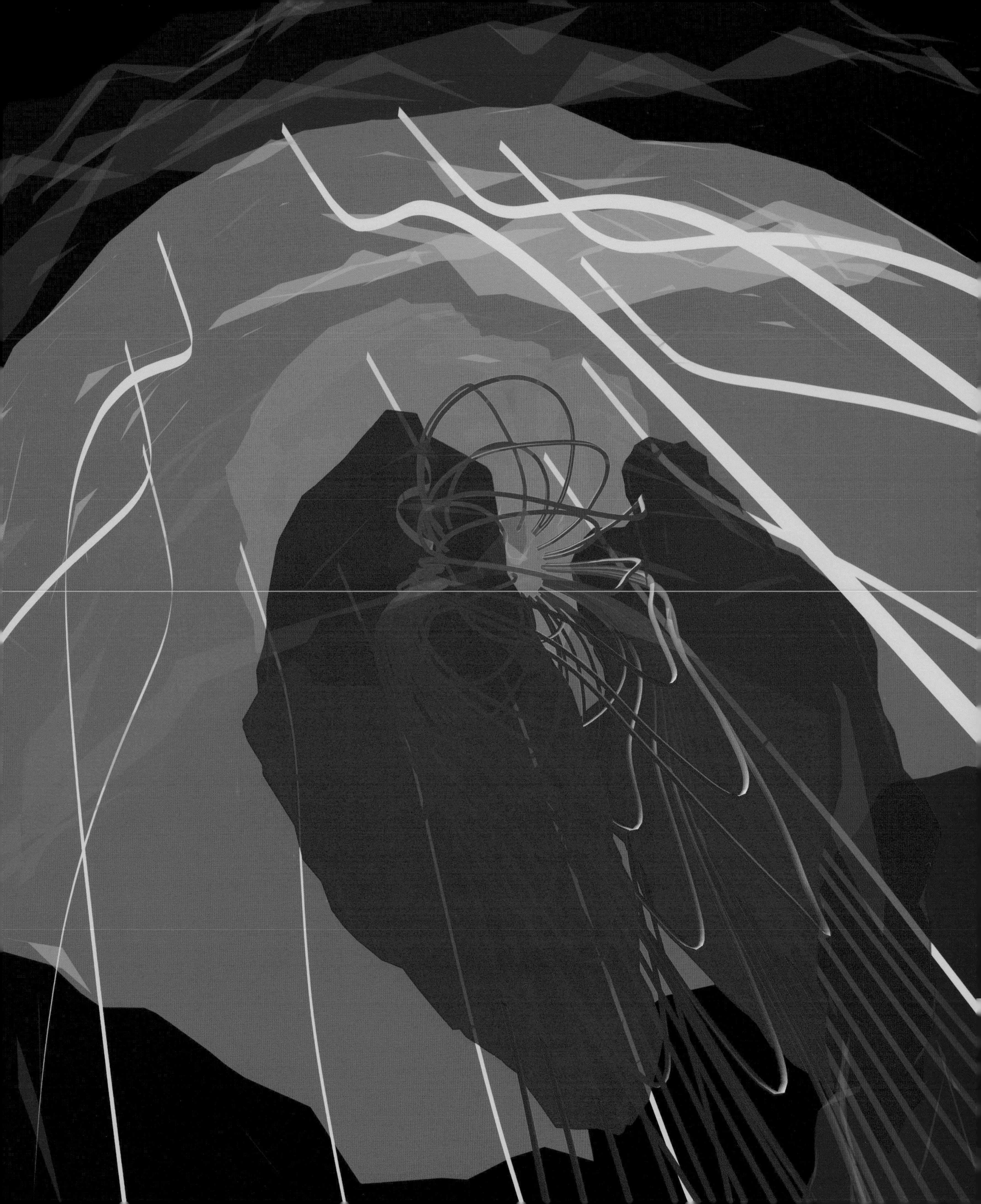

Mysterious moons
startling discoveries on small worlds

Most of us can name the nine planets in our solar system, but what about all the moons? There are at least 124 scattered among various planets. They include a chemically mysterious world with a dense atmosphere, a hellish ball of rock in constant volcanic disarray, and an ice moon that may be hiding an ocean. There are also several dozen captured asteroids that can barely be described as 'moons'.

Jupiter has 58 moons. Of these, 31 are so small they have yet to be assigned names. The surface of Io, the third largest, is in permanent upheaval. The tidal forces generated by Jupiter's gravity are so great that Io's crust can never settle; it broils constantly, and most of it is subject to volcanism. From space, it seems as messy as a molten pizza. The crusts of Ganymede, Jupiter's largest moon, and Callisto, the second largest, are less turbulent, but the patterns frozen into their crusts are fascinating.

Neither these nor most of Jupiter's other moons are suited for life, but there may be an exception: Europa. NASA scientists believe that its crust is a thick layer of ice. Wide-scale fracturing is caused when Jupiter's gravity tears the crust apart at frequent intervals. Certainly the fracture patterns are reminiscent of polar ice floes right here on Earth, and there may be a huge ocean of liquid water, 50 kilometres deep, beneath the ice layer. Quite possibly, it is slightly warm. When it pushes up between fresh ice cracks and freezes into place, it deposits grains of darker material on the surface. Spectral analysis conducted by the Galileo space probe in 1997 suggests, very tentatively, that it may have an organic component.

Saturn's most intriguing moon is Titan. Twice the size of our own, it is the only moon to possess an atmosphere – one so dense we cannot see through it with conventional cameras. Nitrogen is the principal constituent, but it is the presence of water vapour and hydrocarbons (methane, for instance), in a wide range of molecular types, that has excited space scientists. The surface may consist of rock interspersed with liquid hydrocarbon lakes. Therefore, although probably unsuitable for life, Titan may be a 'pre-biological' world where the basic chemical ingredients are in place – albeit not yet in sufficient complexity.

Such fascinating discoveries mean that there is plenty of work in the solar system yet for future generations of robotic space probe.

The icy crust of Europa shows patterns of fracture and refreezing similar to those already observed, by air and satellite, in the Earth's polar regions. Evidence from space probes suggests that a deep, warm-water ocean exists under the ice.

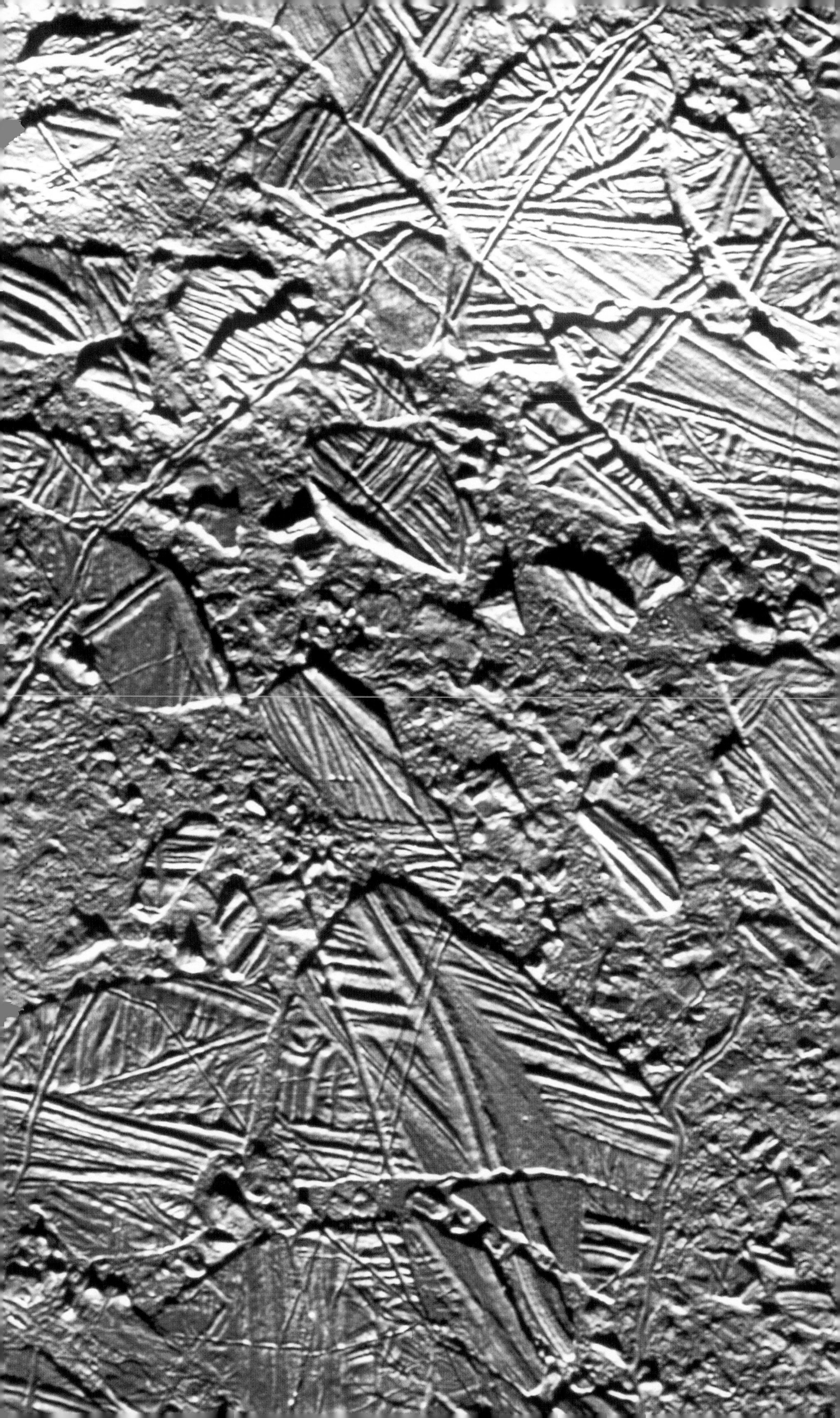

Beneath the glare

the Sun's amazing structures revealed

For generations, we have imagined the Sun as a featureless fireball: impressive, but not particularly structured or complex. Nuclear theory demonstrated that its inner workings must be subtle and dynamic, but it was hard to picture any of these forces until we learned to see past the intense ultra-white glare and into the broiling heart of the beast. Skylab astronauts took X-ray photos in the early 1970s, revealing vast flares and other phenomena, but these were, essentially, just improvements on what ground telescopes were already capable of. Today, high-magnification instruments working at the extreme end of the ultraviolet spectrum manage to cut out extraneous glare. The Sun is revealed as a lumpy old thing, far from the perfect sphere of our imaginings.

In the summer of 2003, astronomers working with the Swedish one-metre Solar Telescope on the Spanish island of La Palma captured extraordinary three-dimensional ultraviolet images of active regions on the Sun's surface. The trick was to look not at the Sun's central regions but rather towards the edge, or limb, of the disc. The effect is similar to staring down at a crowded sidewalk from above and seeing only the tops of the heads of passers-by, then looking towards the end of the block and, with a more oblique perspective, being able to see more of their bodies.

The granular features are evidence of the rapid convection that transports heat up to the surface. In principle, the effect is similar to boiling water on a stove – albeit on a colossal scale. Each of the granules is hundreds of kilometres tall, and covers an area about the size of Texas. Sunspots and smaller dark pores are sunken into the surrounding granulation. The optical colouring of this processed image is arbitrary, a yellowish hue, conforming with how we think the Sun should look.

Our Sun has burned for 5 billion years. It will burn for 5 billion more, after which it will become an unpredictable monster, a 'red giant' prone to sudden nuclear shutdowns and re-ignitions, expansions and contractions, alternating over several millennia. Its wayward mantle of gases will swallow up half the planets of the solar system, including the Earth. All life will be vaporised …

Solar flares are regularly expelled deep into space (below) and these have been widely studied over the last half-century. The image on the right, taken in extreme ultraviolet, shows a new technique for imaging the Sun's actual surface in unprecedented detail.

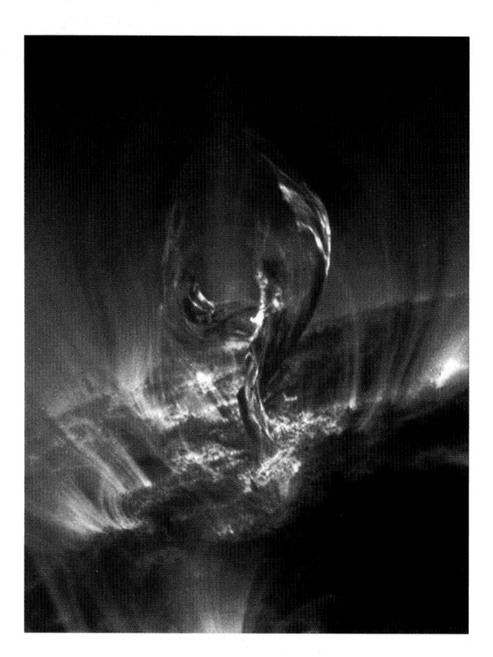

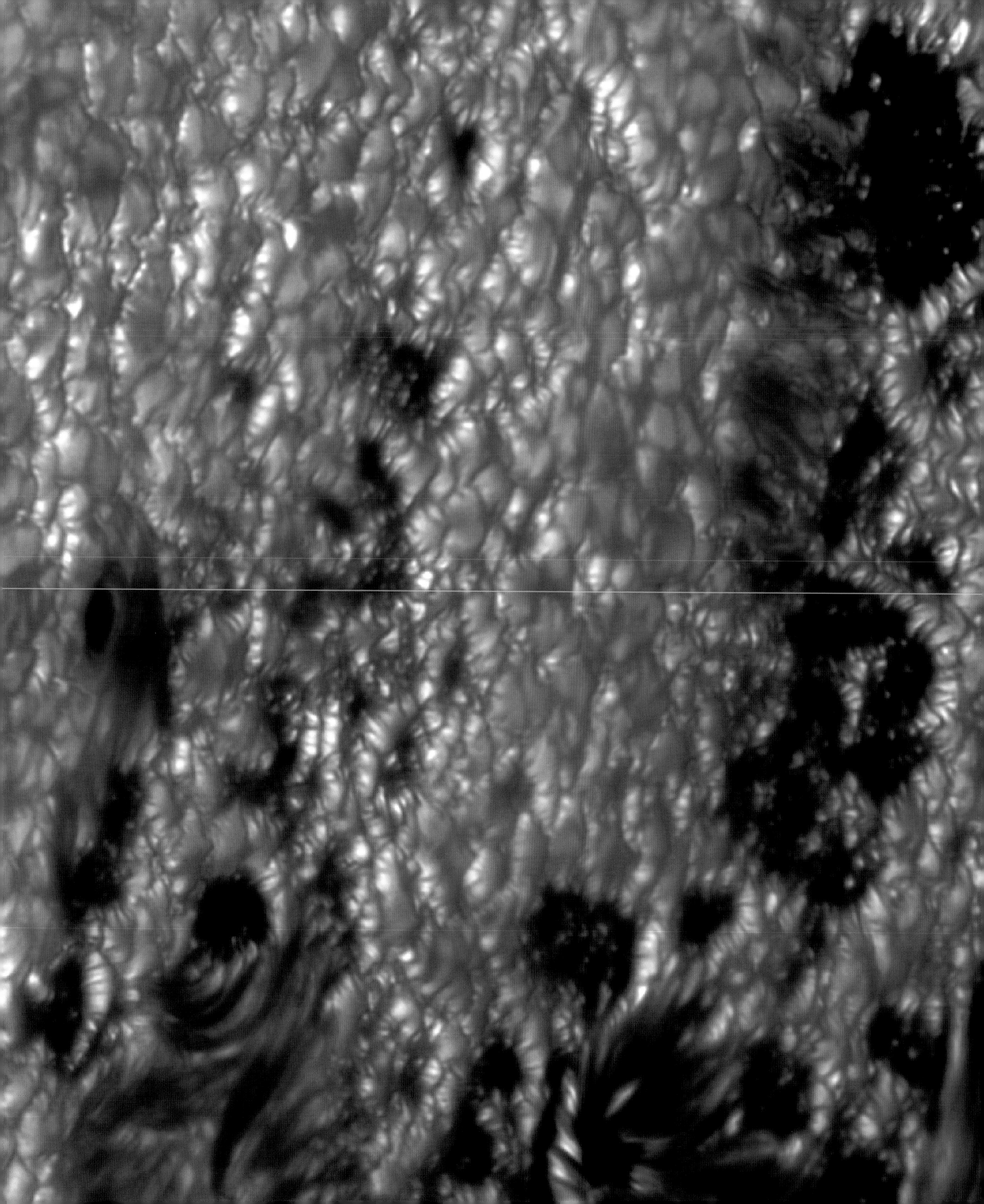

Simulating the spark
modelling star formation

The birth of a star is a dramatic process. First, vast clouds of hydrogen and helium condense from the dead remnants of earlier generations of stars. The clouds drift through interstellar space for countless millennia, pulled about by galactic tidal forces. Somewhere within this cloud, a seemingly insignificant clump of matter coalesces and increases in mass until it is capable of attracting other nearby atoms and molecules. The gravitational forces involved are so delicate at first as to be virtually non-existent. But not quite. Over time, the little clump of matter grows until its gravity becomes strong enough to pull in more material, from distances up to several hundred kilometres. A 'protostar' has formed.

After more millions of years, a large region of the cloud begins to spiral towards the embryonic star, the spin of which forces this region into a disc shape. The protostar's mass increases, and so does its gravity field. Its reach now extends billions of kilometres into space, pulling in yet more matter, building up greater gravity and piling on the pressure at the core. Eventually the pressures and temperatures become so intense that matter can no longer exist in any conventional state: the core ignites in a gigantic thermonuclear reaction, and a new sun explodes into life.

As our Sun sucked up material from its accretion disc, a smaller rival fed on the same disc, gaining mass and building an immense gravitational field – but not quite enough to create the high temperatures and pressures at its core that a true star needs. Had it been able to sweep up hydrogen from the disc for a little longer, it would have burst into light and challenged the Sun for supremacy. A binary star system might have been created, if only the surrounding supplies of dust and gas had not been depleted just as the competition was getting into its stride. Binary systems are not healthy places for planets to live in, nor for life to develop. We should be grateful that the failed competitor settled for becoming Jupiter.

One of a series of frames from a computer simulation, demonstrating gaseous material from a nebula being drawn towards a new star by its growing gravitational influence. Nebulae often comprise debris from older stars that exploded at the end of their life cycles.

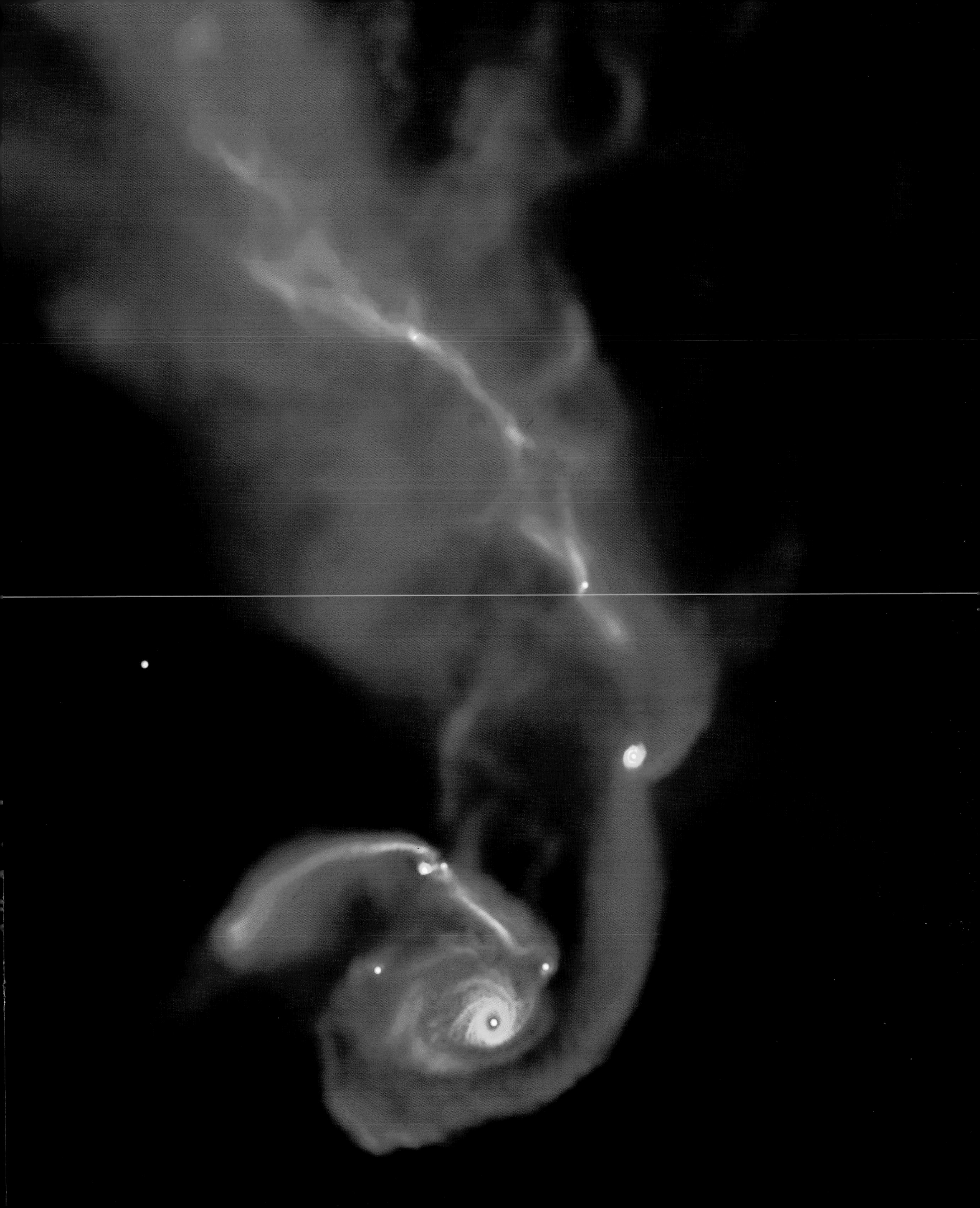

Dying stars
supernova explosions

Large stars die a quick and bright death known as a supernova. When the star's nuclear fuel reactions are exhausted after billions of years' constant fire, the outer layers of hot gas collapse. A massive explosion then blasts those layers into space. Meanwhile, the remains of the star are compressed by the force of the explosion. If the star was sufficiently massive to start with, the core will shrink under the influence of its own gravity yet further, to a pinprick; and a black hole is born.

Supernovae are rarely seen. There have only been six observed in our Galaxy over the last thousand or so years, and sadly none of them happened within the age of modern astronomy. What we today call the Crab Nebula in Taurus is the remnant of a supernova that was visible in the night skies of 1054. The most recent blast was that seen by the famous astronomer Johannes Kepler in 1604.

However, we have seen a reasonably close supernova just beyond our Galaxy. On the evening of 23 February 1987, Canadian astronomer Ian Shelton took a picture of a nearby galaxy, the Large Magellanic Cloud (LMC). He processed the photo and immediately noticed a bright star where none had been seen before. It was a supernova. Ground telescopes around the world soon swivelled to catch the dying glory of SN1987A, and seven years later the Hubble Space Telescope photographed the still very fresh expanding gas blasts from the star. This typical disaster was not just an ending for one star. Supernovae are the mechanisms by which spare hydrogen fuel, and chemical elements processed and created inside stars, are scattered into space when those stars reach the end of their life cycles. Clouds from old supernova remnants eventually coalesce, drawn together by the gravity of their local galaxy, and new stars condense out of them. If our universe contains only a fixed amount of matter and energy, this process must eventually wind down.

This computer simulation investigates the likely characteristics of supernova SN1987A. It analyses the interior of the star and its rapidly expanding outer shock wave of gases, 27 minutes after the catastrophic collapse of the core and its subsequent detonation.

An analysis of the turbulent convection currents inside a star during a supernova explosion. Some of the data was derived from observations of an actual supernova, recorded by astronomers in 1987. A vast shock wave is emanating from it after the core collapse and subsequent explosion.

Cosmic monsters
the terrible power of black holes

Black holes are the ultimate challenge for observational science, because they are objects which, by their very nature, can never be observed. When some stars die, they explode with such force that material is pushed inwards as well as exploding outwards. The inward-rushing mass becomes so dense it collapses to a pinprick under its own gravity. From this point, any further stray material – gas clouds, even entire nearby stars that come too close – is sucked in, increasing its mass until it suddenly blinks out of the realm of normal time and space, becoming an infinitesimally tiny pinprick of infinitely condensed matter with a truly monstrous gravitational field. Nothing can escape this field – not even light; hence the term 'black hole'. The centres of galaxies are denser than their perimeters, and black holes are more likely to form there because of the relative abundance of material to feed upon. Some scientists even suggest that supermassive holes are a necessary evil: a centre of mass around which healthy galaxies can coalesce.

The 'other side' of a black hole may be a 'white hole' in another universe. A 'big bang' emerges from a dimensionless point of infinitely compressed matter; black hole singularities also are dimensionless points of infinitely compressed matter. The two singularities, 'black' and 'bang', are like mirror reflections of each other on either side of a space-time boundary. Perhaps black holes created from a few thousand dead suns in one universe can spawn trillions upon trillions of suns in another.

This large image is a computer simulation of gravity waves rippling back and forth immediately after a black hole has swallowed a stray star. The surrounding zone of space-time is distorted in rhythmic pulses. No pictorial representation can make sense of the actual 'shape' of such a phenomenon, but we can sense some of the characteristics of a black hole using familiar optical light. The smaller image yields compelling evidence for a supermassive black hole at the heart of galaxy M84. It shows the spectrum of visible light frequencies emitted from a fast-spinning disc of glowing gas and other material which has come within the black hole's baleful influence, but has not yet quite been sucked in. The gas is being spun around so fast, the wavelengths of its emitted light appear from our point of view to be stretched into the long red end of the spectrum or compressed into the short blue end, depending on whether the material is rushing away from us or hurtling in our direction. What would have been a neat spectrum, not unlike the rainbow made by a glass prism, is crazily distorted. A normal spectrum should appear as a row of vertical strips, each of a different colour, and ranged in order from short to long wavelengths. Here, the jagged shape of the spectrum demonstrates the black hole's distorting effect on light. It is not a 'picture' in any sense that we might normally use that term, yet the image does surely convey the power of this terrifying vortex. Analysis of the data suggests that M84's black hole may have swallowed the mass equivalent of 300 million suns since its birth.

This mathematical model renders space-time as a geometric surface contorted by a black hole. The smaller image, captured by the Hubble Space Telescope, shows wild distortions in the light spectrum of a doomed star, caused by its extremely rapid rotation around a black hole.

Seeing the generation gap
ultraviolet images of galaxies

When seen through a conventional low-powered optical telescope, most of the stars in any given galaxy look to our eyes, let's face it, uniformly bright white. This does not detract from the beauty of a galactic swirl, but we have become used to the subtly enhanced images from Hubble and other instruments, where different colours of light are electronically exaggerated so that we can more easily see the gradations. Although these are properly called 'false-colour' views, they are still read as broadly naturalistic. We accept that the universe is being presented, as it were, in its best possible light.

This view of galaxy M33, a member of our local group, takes enhancement to an extreme degree, differentiating between visible (red) and ultraviolet (blue) sources. The intervention of the image-making process is so extreme that we would no longer describe the picture as natural-looking. But it is still beautiful. The strong ultraviolet sources are associated with younger stars – a generational difference that is not so apparent in ordinary light.

The universe is running out of raw materials to make new stars – or, at least, it is running out of material of sufficient density. The net amount of material remains fixed, but energy and matter are gradually becoming more diffuse. Many galaxies are still spawning fresh stars from gas clouds left over when previous stars exploded, but it may be that galaxies themselves are no longer being made. The universe is middle-aged and past its reproductive prime, and not enough new stars are being created overall to replace the ones that die. Somewhere around 5 billion years from now, when our own Sun approaches the end of its life, the universe may only be half as bright with stars as it is today.

All stars and galaxies must eventually burn themselves out. Throughout much of the twentieth century, cosmologists have argued that the universe is destined to suffer 'heat death', the diffusion of all its heat energy until nothing anywhere is more than a few degrees above absolute zero. The only way out of this terminal decline would be to keep pumping new energy into the system from within – energy hidden from our instruments in the vacuum of space. Modern cosmologists are exploring just such a possibility.

This view of galaxy M33 identifies strong ultraviolet light from the youngest, hottest stars. The concentration of young stars in a galaxy's core is probably due to the greater availability of gas in that region, and the more frequent recycling of old stars into new.

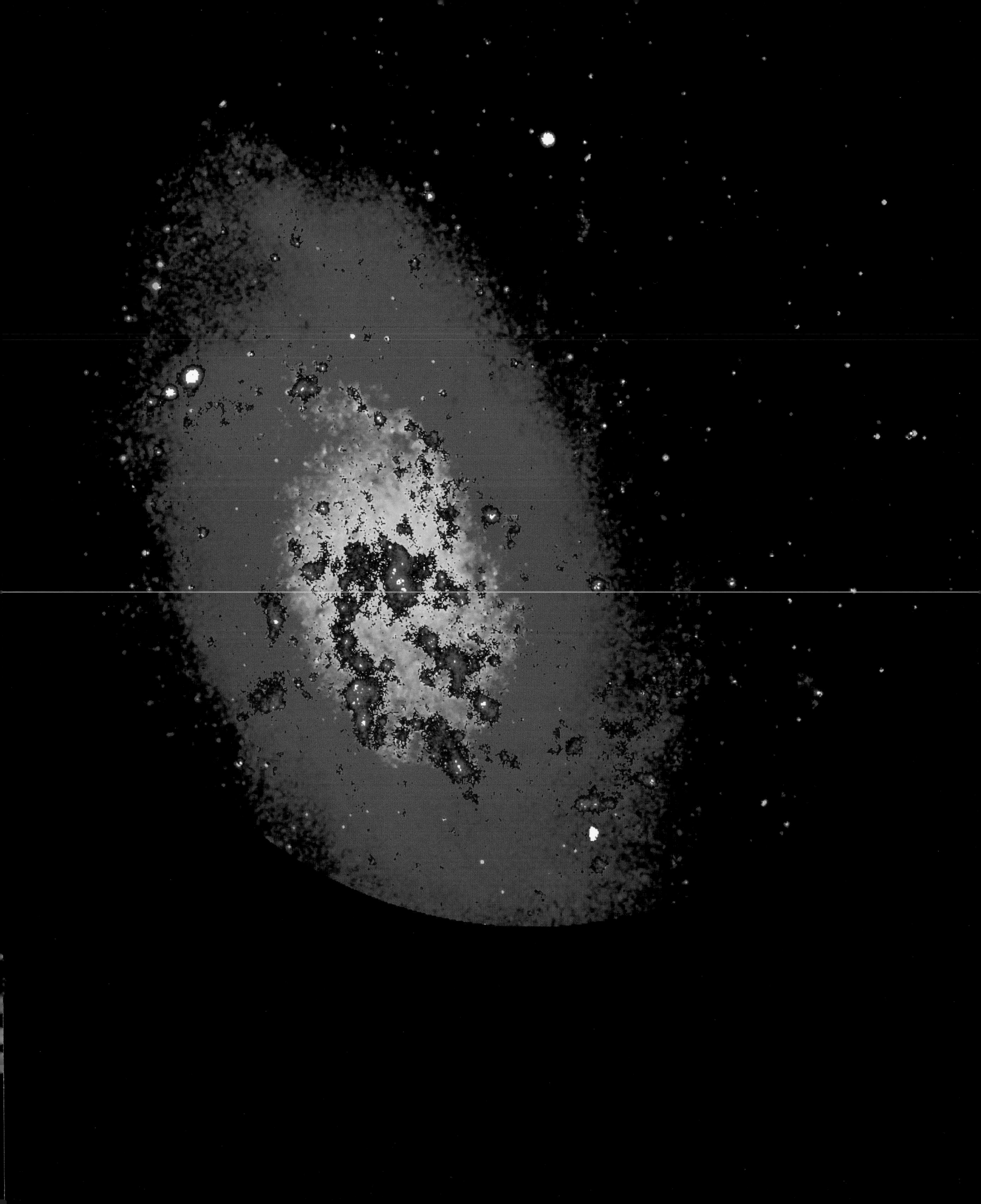

A trillion suns

the mysterious energy of gamma ray bursts

In the late 1960s, the top-secret Vela class of military spy satellites was launched to detect gamma rays. The intention was to pick up traces of Soviet nuclear experiments and underground bomb tests. In a 1973 report, Ray Klebesadel, Ian Strong and Roy Olson of the Los Alamos Scientific Laboratory announced that Vela detectors had indeed picked up high-energy gamma ray flashes – but most of the signals came from deep space, rather than Earth. It turned out to be a momentous discovery. Gamma ray 'bursts' lasted from fractions of a second to several minutes, popping off like brilliant cosmic flashbulbs from unexpected directions and then fading away, much to the initial consternation of the military. If only they could have imagined how insignificant all the atomic bombs ever created would seem in comparison to these celestial monsters, which release more energy in a few seconds than the Sun in its entire 10 billion-year lifetime. As far as we know, they are the most violent and powerful events in the cosmos apart from the Big Bang itself.

During the 1990s, the Burst and Transient Source Experiment (BATSE), an instrument aboard NASA's orbiting Compton Gamma Ray Observatory, confirmed that bursts occur on a daily basis, at random times and from random directions. A map has been created that shows a history of recent bursts distributed all over the sky. However, when other instruments peer into the regions where those bursts have flared briefly and then faded, they cannot find any specific object associated with them – and there is no evidence, yet, that bursts can eventually occur in the same place twice in succession. This suggests that each burst originates from some form of cataclysmic explosion, never to be repeated in that location in space. So far, it appears that all bursts originate from outside our Galaxy, but this theory is based mainly on the premise that such astoundingly powerful sources of energy could not possibly be any closer to us.

On 28 February 1997, an Italian/Dutch satellite known as BeppoSAX detected X-rays lingering a full eight hours after a gamma ray burst: an unprecedented opportunity to bring ground-based telescopes on to the target before all the locational evidence vanished. Astronomers identified a dim, distant galaxy in the same point of the sky (within the constellation of Orion). Was this proof, at last, that bursts must be associated with galaxies? Or was it merely a coincidental overlap? Astronomers are now reasonably confident that a link has been established. Three days later, a second scan (the right-hand image) showed almost no sign of the burst.

These two scans, taken just three days apart by a Dutch/Italian satellite, show a gamma ray burster flaring briefly, liberating immense energies before fading away. It is unlikely that a burst will be observed again at that location.

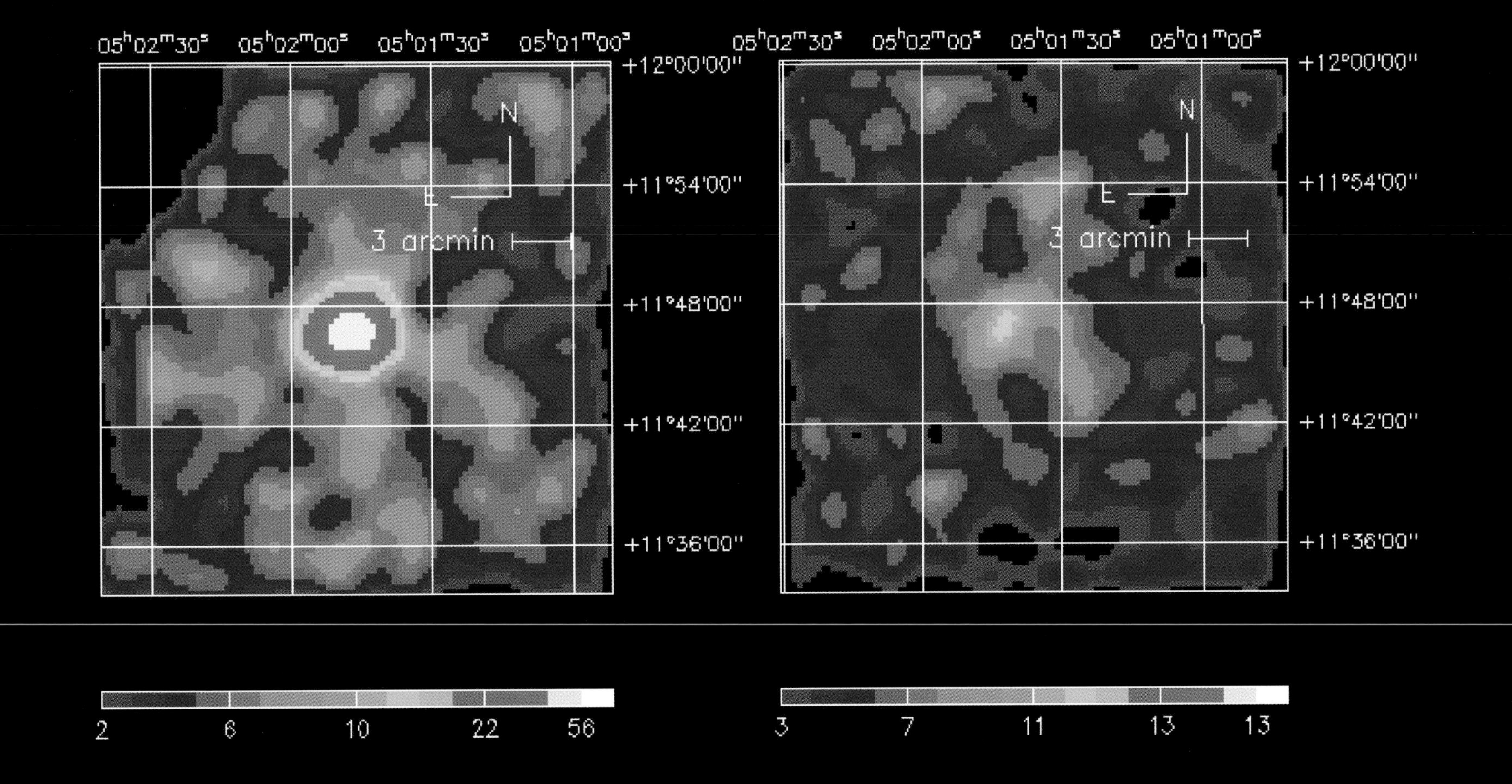
05h02m30s
05h02m00s
05h01m30s
05h01m00s
+12°00'00"
+11°54'00"
+11°48'00"
+11°42'00"
+11°36'00"
N
E
3 arcmin
2
6
10
22
56
05h02m30s
05h02m00s
05h01m30s
05h01m00s
+12°00'00"
+11°54'00"
+11°48'00"
+11°42'00"
+11°36'00"
N
E
3 arcmin
3
7
11
13
13

The loud universe

making pictures with radio waves

In 1932, the American astronomer Karl Jansky detected radio signals coming from far beyond the Earth's atmosphere; in fact they stemmed from the centre of our Galaxy. A galaxy comprises not only stars, but also the vast swathes of tenuous gas that exist between the stars, and from which stars are born. Sometimes these gas clouds are sufficiently energised by nearby stars to give off visible light, but radio energies are a more reliable way of detecting the thin, cold clouds of material drifting through the immense expanses of a galaxy's outer regions, where stars are sparse. The smaller image on this page shows galaxy Centaurus A, located about 16 million light years from us. An optical image might resolve the galaxy into its constituent stars, but this image maps, instead, the overall contours of its radio energy, generated in large part by its gas clouds.

The extraordinary picture to the right was also made with radio waves, collected by the Very Large Array (VLA) radio telescope in New Mexico. It shows a quasar ('quasi-stellar' object). Quasars generate electromagnetic energy equivalent to 100 times the output of a normal mature galaxy. They sometimes appear as star-like pinpricks in an optical telescope, but they are more usually identified by powerful emissions in the radio realm of the spectrum, which are given out by energetic gas particles in their vicinity. In this example, the quasar itself is at the top of the page, and the vast bolus of gas which it has ejected is visible at the bottom. Current theory suggests that quasars are connected with supermassive black holes, but no one is absolutely sure of the link.

Quasars seem to be moving away from us at a rapid rate. This means that any electromagnetic light we observe as originating from them is greatly stretched ('red-shifted') into the far red end of the spectrum. The effect is analogous to the whistle of a train seeming to lower in pitch as the train hurtles past us, because the wavelengths of the sound are stretched out. All galaxies outside our own local cluster exhibit red shift to some degree, due to the overall expansion of the universe and the acceleration of galaxies away from each other. But quasars are much more compact than galaxies, and their red shift is particularly extreme. The further away an object is from us, the faster it is speeding away, and the greater its red shift appears to us. Judging by their red shifts, quasars must be the most distant objects we have ever observed. Their light has taken more than 10 billion years to reach us, and if so we must be witnessing phenomena from a time when the universe itself was very young.

However, we cannot yet be conclusively sure that quasars are indeed as vastly distant as theory suggests. It is possible that many exist within our own Galaxy, and their tremendous energies and red-shift phenomena may be the products of some process we have yet to understand.

A radio map of quasar 1007+417, made by the Very Large Array telescope in New Mexico, reveals extended streams and lobes of gas being emitted from the heart of the quasar (top). The image below shows radio contours of galaxy Centaurus A, 16 million light years away.

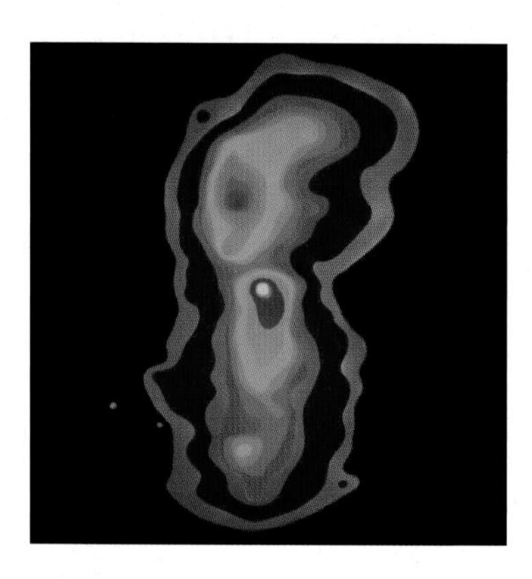

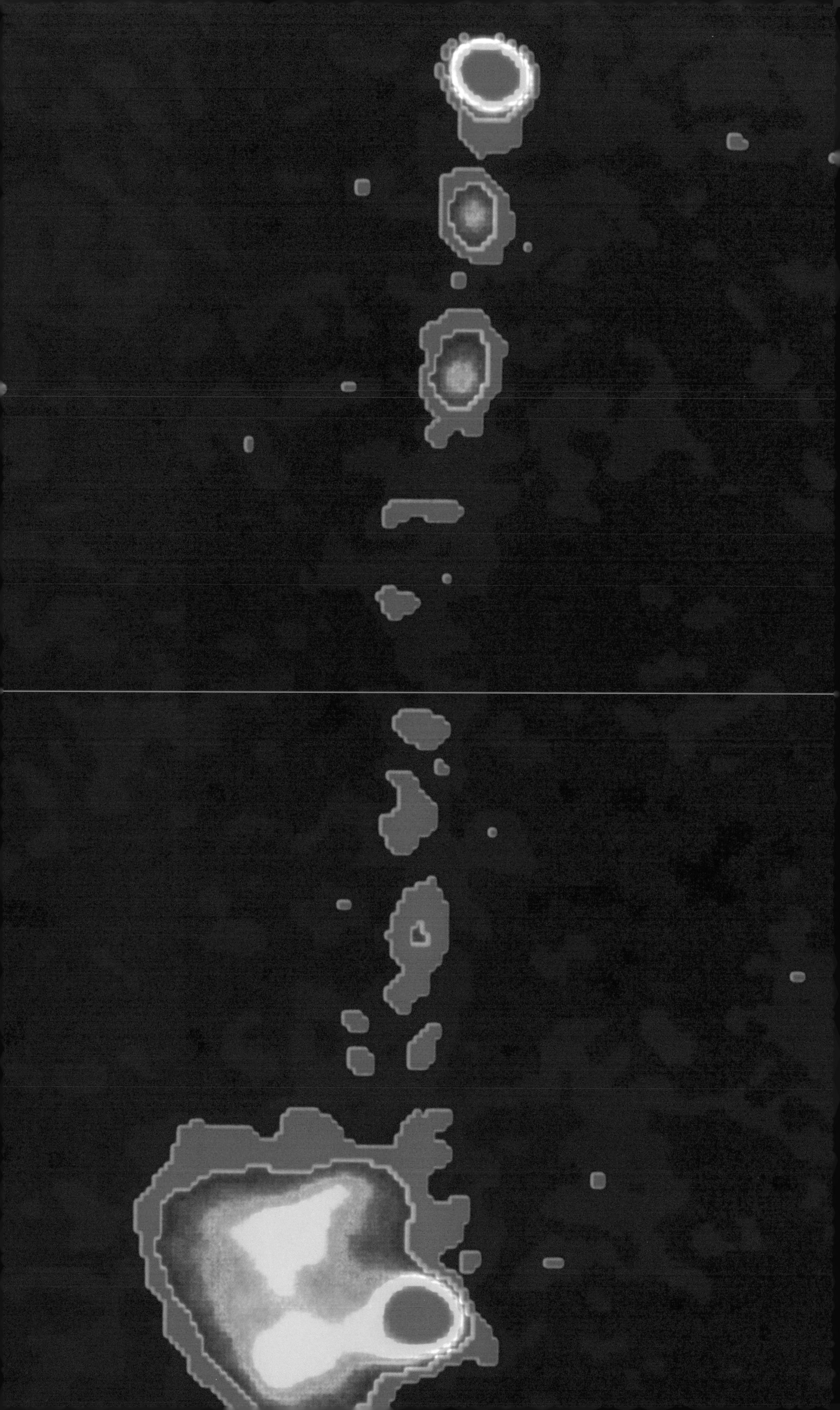

The echo of creation

the microwave residue of the Big Bang

How beguiling this image is: a map showing the uneven distribution of microwave background radiation in the night sky, the almost unimaginably faint reverberations of the Big Bang more than 13 billion years ago. It is a cosmic 'fossil' from the dawn of time.

In 1929, the astronomer Edwin Hubble demonstrated that the universe is expanding in all directions. The further away a cosmic energy source is from us, the faster we observe it receding, and its light stretches into longer wavelengths at the red end of the spectrum ('red-shifts'). The microwave background in this image is essentially gamma radiation from the Big Bang that has been red-shifted to the other end of the spectrum by the subsequent expansion of the universe.

The raw data was gathered by the Cosmic Background Explorer satellite (COBE) and the resulting interpretations made headline news around the world in 1992, proclaimed by excitable journalists as 'the fingerprint of God'. We might be tempted to look up at the stars one night and, in our mind's eye, superimpose this map. In fact COBE's persuasive picture of the echo of creation is not so straightforward. Far from being instantaneous snapshots, COBE's measurements were gathered over more than a year. In that time, the Earth made a complete revolution of the Sun and, accordingly, the perspective of the night sky from COBE's point of view was substantially shifted. Meanwhile, COBE flew a polar orbit, oriented sideways to its direction of travel with its delicate instruments permanently pointed away from the Sun. The spacecraft dared not swivel its gaze and had to wait, month after month, for those parts of the sky that were not obscured by the blinding glare of the Sun to come into view. The neat and tidy map that emerged from these ever-shifting perspectives had to be coaxed out of a morass of complex data.

COBE revealed microwave aftershocks left over from a time when stars and galaxies, and even matter itself, had not yet formed. The scattering of signals in the map, billions of light years across, are the weak remnants of energies that were once compressed in the first super-hot pinprick of creation. Without these asymmetries in the first moments of the Big Bang, the universe would have expanded with coldly beautiful but utterly sterile evenness, perhaps eventually dissipating into nullity. Around those messy, random imperfections in the first moment of creation, matter condensed into the lumpy and diverse universe we know today.

In February 2003, the Wilkinson Microwave Anisotropy Probe (W-MAP) resolved the details further, yielding a wealth of information that slotted neatly into the Big Bang theory: the universe is 13.7 billion years old, and the microwave energy dates from 379,000 years after the Big Bang.

COBE made headlines in 1992, but this map was made much more recently, in 2003. It greatly refines the cosmic temperature map: the universe is 13.7 billion years old, and the microwave energy dates from 379,000 years after the Big Bang.

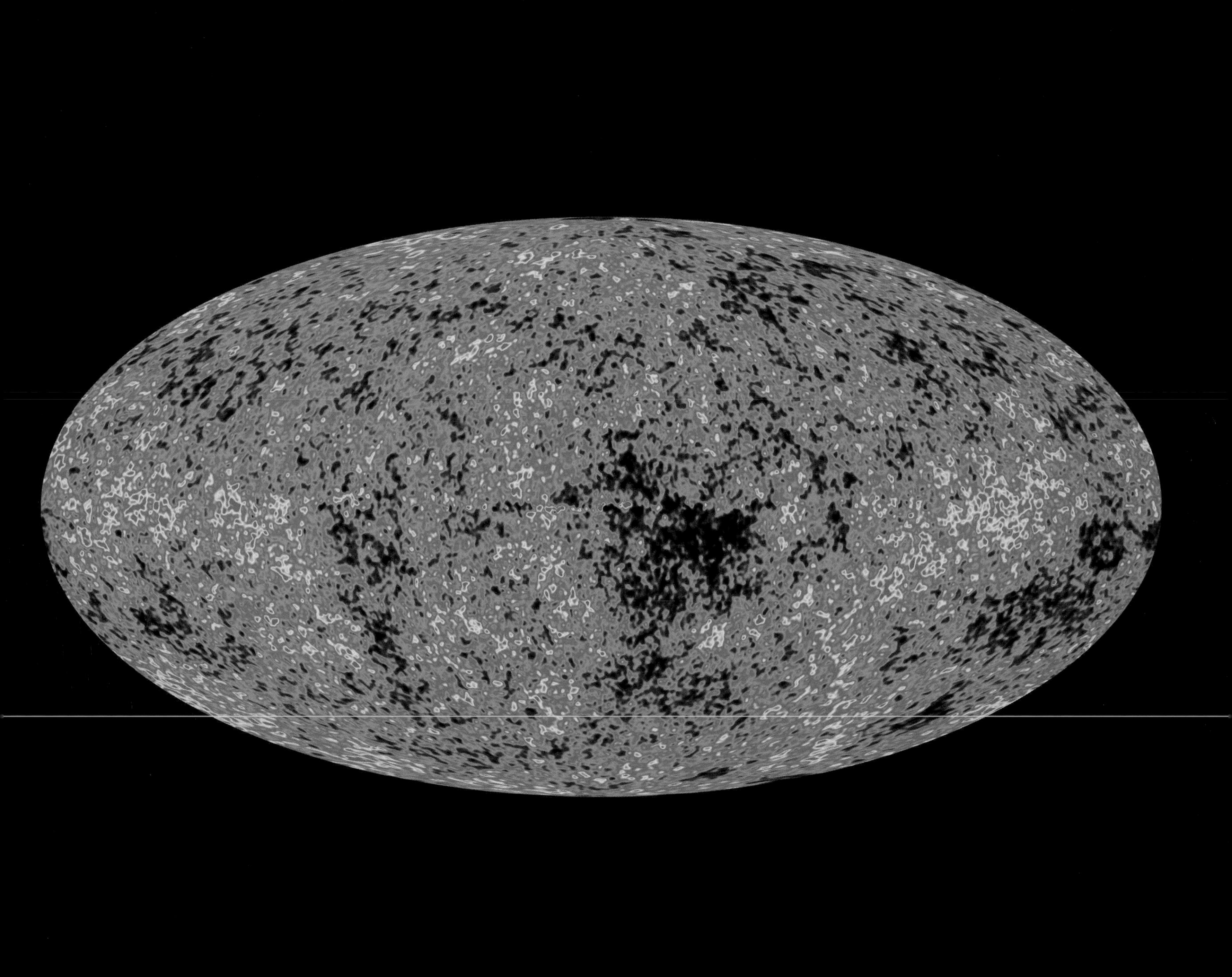

Most of everything is missing
the hunt for dark matter & energy

As far back as the 1930s, astronomers recognised that galaxies in large clusters drift around each other under the influence of their collective gravity, but faster than their apparent masses would dictate. The gravity is stronger than a simple tally of the galaxies alone can explain. In the 1970s, instruments measured the rotation of individual galaxies too, by comparing the wavelength shift of light from their inner cores and outer rims. The rotations showed that galaxies contain a great deal more mass – 'dark matter' – than can be explained by the presence of stars and gas clouds. One theory suggests that dark matter is made of WIMPs, subatomic 'weakly interacting massive particles' which possess mass, and therefore create gravitational effects, but which otherwise hardly interact at all with normal matter, making them very hard to detect.

Light from distant galaxies is bent by the gravity of closer foreground galaxies – gravity generated mainly by dark matter. In March 2000, a French team at the Canada-France-Hawaii Telescope analysed the shapes of 200,000 faint (extremely distant) galaxies in a particular region of the sky. They compared the distorted shapes against the normal shapes of closer galaxies, and over two years built a map inferring dark matter distribution in that region. The filaments of dark matter, shown here in red, are of course invisible, but the distorted galaxy images tend to be stretched parallel to the filaments. This allowed the filaments' possible distribution to be calculated.

In 1998 the rate of expansion of the universe was found to be accelerating. Some kind of 'dark energy' is pushing galaxies away from each other, essentially like an anti-gravity force – but no one can yet explain what that energy is, or where it comes from. Dark matter and dark energy constitute most of what exists in the universe. The billions upon billions of planets, stars and galaxies, and the vastly powerful energies and radiations that create and destroy them, account for barely one per cent of the totality. Almost all of everything is invisible to us in the most fundamental sense of that word.

This map of dark matter distribution, highlighted in red, has been extrapolated from astronomical observations of the light from distant galaxies. Strands of dark matter reaching across the universe may be responsible for shaping the overall patterns and concentrations of galaxies.

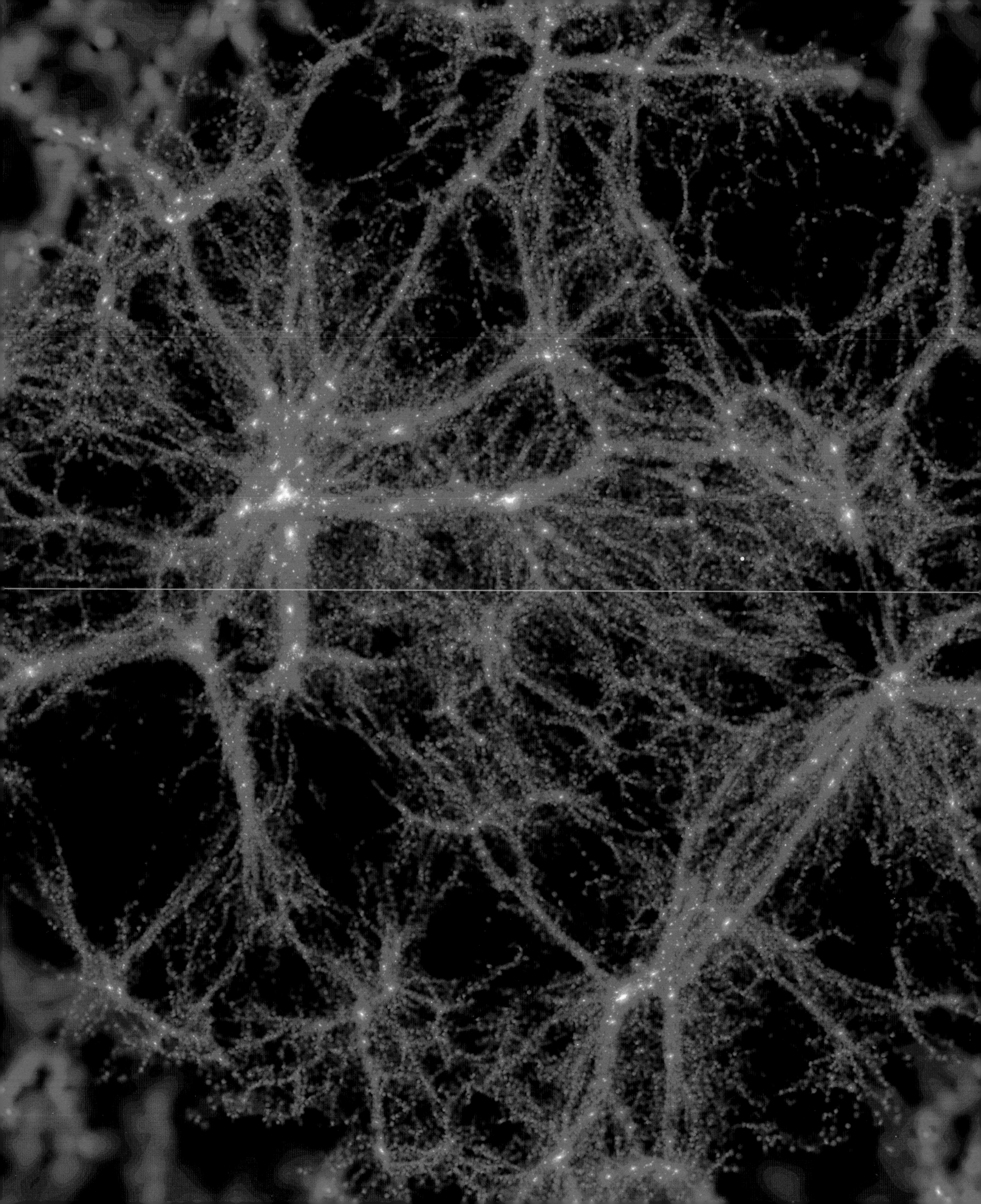

Beyond light

7

The deeper meanings of equations can now be explored by applying them many millions of times in high-speed computers. In the process, we seem to be unveiling fundamental underlying patterns of nature. It could be argued that these results are discovered rather than invented. Is mathematics just a human creation, or is it somehow embedded in the fabric of reality, independent of human culture?

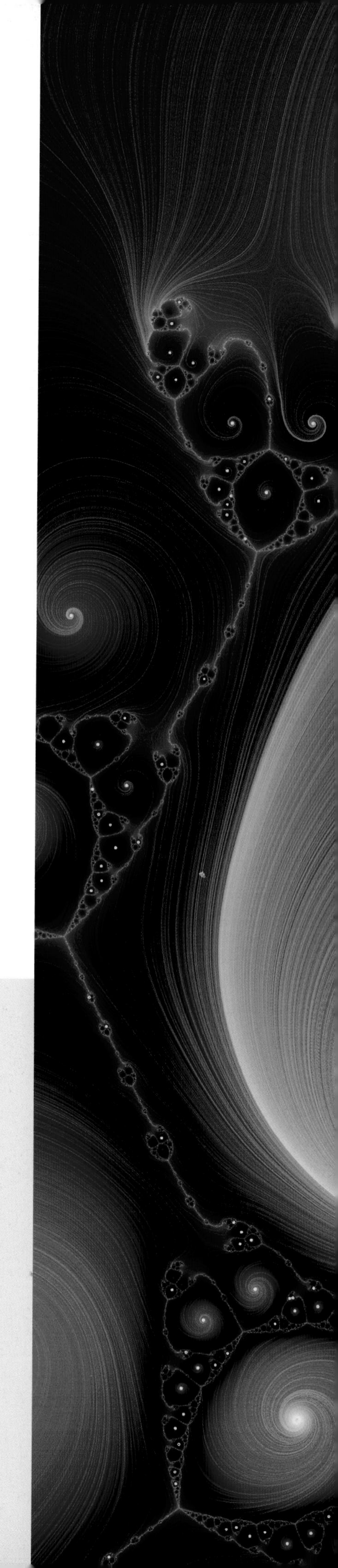

The golden ratio is mathematics embedded in nature?

The Fibonacci series, named after the thirteenth-century Italian mathematician who explored some of its wider possibilities, simply involves taking two consecutive numbers and adding them up to get a third. You then add the second and third to get the fourth, and then the third and fourth to get the fifth, and so on:

1 + 1 = 2   1 + 2 = 3   2 + 3 = 5   3 + 5 = 8   5 + 8 = 13   8 + 13 = 21   13 + 21 = 34

The ratio between successive numbers in the series never settles at a specific value, but the more numbers are fed into the process, the closer the ratio tends towards 0.61803399…, an 'irrational' number, somewhat like pi (in fact it is often nicknamed 'phi') that cannot ever be expressed as an exact fraction. Known as the 'golden section', and often rounded off to 8:13, it has intrigued artists for generations, especially when rectangles whose sides concur with the golden ratio are used as the basis for dividing the composition of a painting. When golden ratio rectangles are stacked around each other in an ever-expanding series, and a smooth curve links the corners of the rectangles, a perfect spiral results.

Fibonacci spirals (along with many other offshoots of the Fibonacci phenomenon) are found widely in nature. A nautilus shell sliced in half reveals a stunningly precise example. A sunflower head is formed of two opposite sets of spirals, 21 going in one direction, and 34 in the other. A pine cone? 5 and 8. A pineapple? 8 and 13. Fibonacci spirals solve what mathematicians call a 'packing' problem. They are an efficient way of clustering multiple, similarly shaped components into a small area, so that no potential growing space is wasted; and they also ensure that no one component is impeded or blocked (from sunlight, for instance) by the others. It takes a very dry intellect not to be intrigued by the fact that these and other mathematical abstractions seem to be embedded in the structures of the real world.

PRECEDING PAGES
This beautiful pattern is derived from a simple mathematical equation called a 'Julia set'. Modern computers allow such equations to be calculated millions of times in succession, so that their deeper implications can be studied as never before.

FACING AND BELOW
A nautilus shell, sliced to reveal its cross-section, reveals an almost flawless example of mathematical precision in nature. Its spiral matches the Fibonacci series. Fibonacci patterns can easily be found in many flowers and plants, as in the sunflower below.

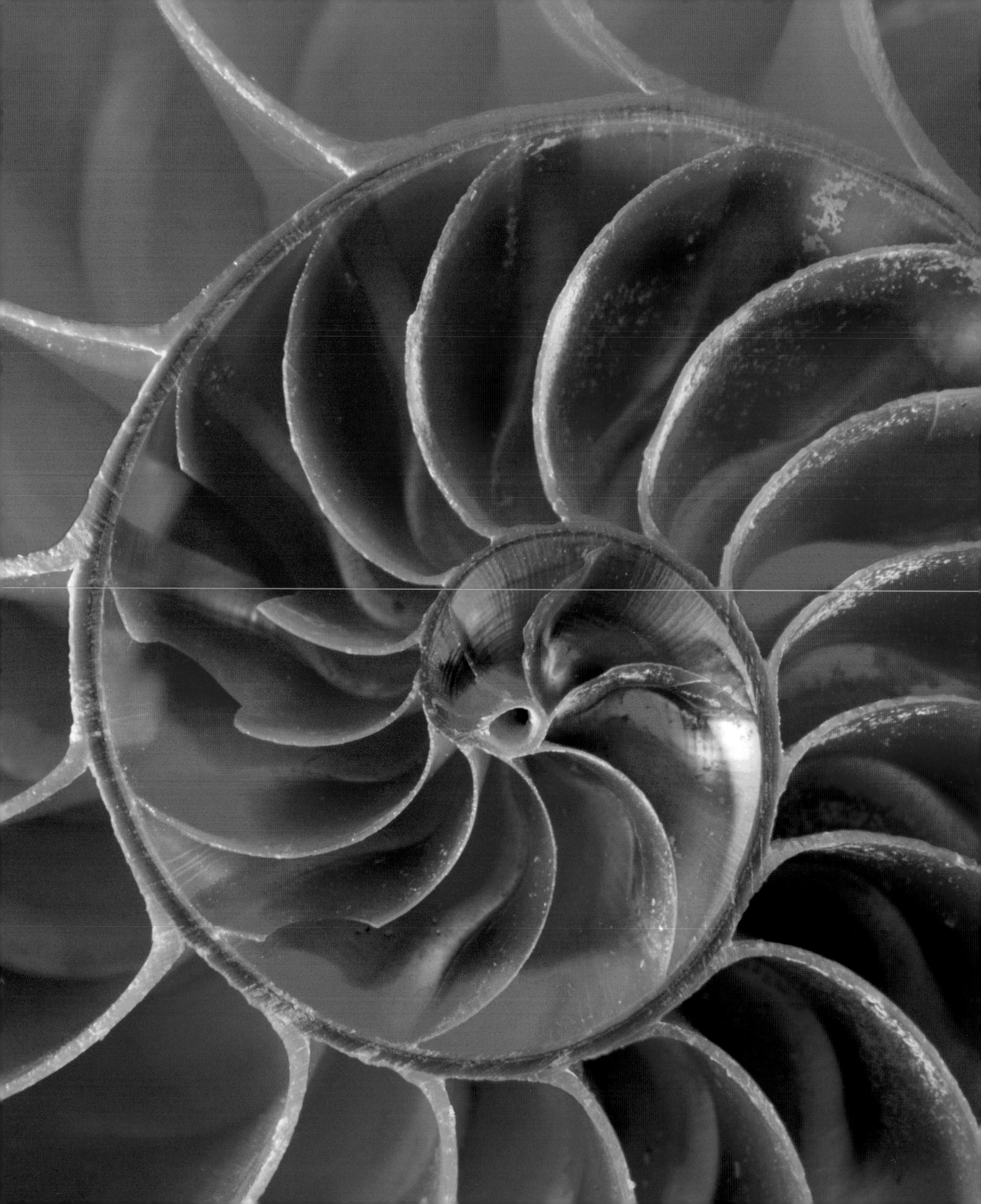

Mathematical telescopes

computers reveal the patterns in equations

Equations can appear deceptively simple when calculated a few dozen times. However, over the last half-century, computers have served as mathematical 'telescopes', repeating calculations more often in a few milliseconds than any human could in a lifetime with paper and pencil. When the results of each calculation are plotted as a point in a two- or three-dimensional grid, patterns and shapes often appear whose characteristics we had never anticipated. The more points we can calculate, the more detailed the emerging picture becomes. Since the 1980s and the advent of high-resolution colour computer graphics, three-dimensional pictures of mathematical functions have reached such a degree of detail that they seem almost as convincing as objects in the real world. Just as scientists plan more powerful accelerators to investigate the limits of the subatomic realm, so mathematicians now look forward to even faster supercomputers that can capture images from the farthest extremes of the mathematical universe.

Are these mathematical abstractions, or true reflections of nature? Fractals, for instance, are simple equations whose main characteristic is that the result from each calculation is fed into the next one as input: a feedback loop whose initial components can be extremely simple, yet with results that can be astounding if the calculation cycle is repeated sufficiently often. The other characteristic of a fractal is self-similarity at all scales. When a tiny part of a fractal image is magnified, it reveals a sub-pattern that is just like the larger one, and so on, ad infinitum. It appears that nature exhibits similar fractals. Map-makers are familiar with the 'coastline conundrum'. If an outline of a coast is presented on a map, but with no scale of distances attached, it is literally impossible to determine from the shape alone whether the map represents several kilometres of coast or just a few metres. The mineral, biological and meteorological structures of the real world are full of self-similar fractal patterns where the sub-structures emulate the shape of the grander architecture.

Abstract mathematical fractals can be increased or decreased in scale indefinitely. In nature, the physical limit of self-similarity may be at the subatomic scale – although some current research suggests that the quantum behaviour of particles contains its own fractal patterns too. It may be that all of creation is organised in a hierarchy of nested self-similar fractals, from electron quantum waves to supergalactic clusters.

The result from one cycle of a fractal equation becomes the value fed into the next cycle, and so on, for as many cycles as a computer can calculate. Seemingly insignificant variations to the values of the first equation produce vastly different results later in the chain.

The turbulent world
the complexity of fluids in motion

The convoluted flow of water over pebbles looks random and unstructured, but it is in fact completely ordered and self-consistent. A hurricane on the other side of the world might seem the epitome of chaos, but it is not. Over time, any flowing stream must eventually feed into a hurricane somewhere. Its waters will evaporate on a sunny day, altering the humidity of the local atmosphere; its cool temperatures will set up convection currents in the air, and they in turn will affect a larger breeze somewhere else, which may eventually contribute to building a hurricane – and all of this happens in ways that are mathematically comprehensible to us. Turbulence theory, gas laws, thermodynamics, convection, conduction, viscosity, laminar flow, vortex analysis, and on and on; there seems to be a branch of mathematical physics to interpret most kinds of environmental phenomena.

Admittedly, we cannot calculate in advance the exact behaviour of something so complex as a hurricane, because we could never hope to gather enough individual pieces of data to feed into our equations. We would need to know the strengths and directions of every last puff of wind on the entire planet, plus all the temperature and humidity variations, and much else besides. As any meteorologist will verify, our weather forecasts are merely approximations that become increasingly less reliable the further ahead we try to look. Small unknowns have tremendous ramifications further down the line. We cannot even predict something so minor as the swirl of water going down a plughole without knowing all about the soapiness of the dirty bathwater, the dying ripples left by the bather as she gets out of the bath, and even the snags of chipped chromium plating on the drain itself.

Nevertheless, we can understand these phenomena through simulation. Rather than chasing around for limited data in the real world, we can fake a fuller set of numbers from our past experience of what the world usually looks like when we do get a chance to measure it. We can load a computer with millions of points, or 'cells', of imaginary water, each holding realistic values of pressure, temperature and velocity. Then we calculate how they interact in a simulated bathtub, again made of data points that are quite like a real bathtub. The imaginary water will now behave (on screen, at least) just as though it were real water disappearing down a real drain. The more the computerised bathwater looks like what we would expect real bathwater to look like, the more we tend to believe that mathematics is not just a way of looking at nature but is embedded somehow into the deepest fabric of reality.

The large image here shows, in extreme close-up, a film of detergent swirling gently across the surface of a soap bubble. The smaller image is a raging cloud system in Jupiter's atmosphere. If these pictures were presented side by side, in monochrome, and with no clues as to scale, we would have difficulty telling them apart. Incredibly similar rules of turbulence define them both, despite the vast differences of scale and power.

The underlying patterns of a storm in Jupiter's atmosphere (below) and the swirls of detergent and water on a soap bubble (right) are essentially the same. Turbulence mathematics seems to be able to describe a wide range of natural phenomena at all scales.

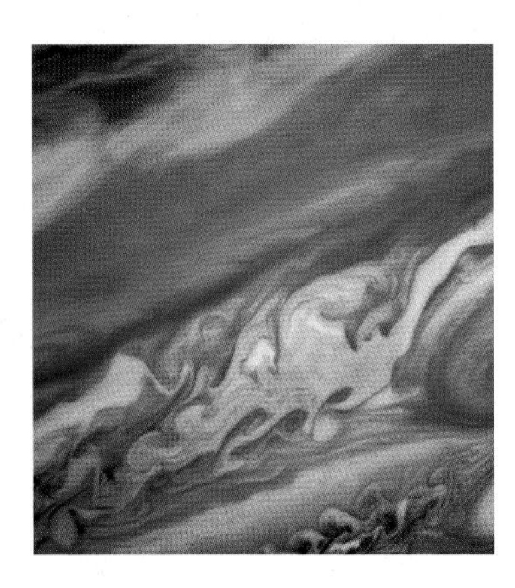

Ideal objects

are mathematical abstract shapes 'real'?

Plato was born in Athens in 427 BC to an aristocratic family. He believed that our sensory experience of the world is merely the dimmest impression of a finer reality. We never see perfect geometrical shapes in the physical world; even the finest mineral crystals contain flaws, and the best human constructions are mere approximations of neatness and symmetry. However, pure shapes do exist in a higher realm, Plato believed. He identified certain forms as having special significance, particularly solids created by folding flat shapes until they enclose a volume completely. Three equilateral triangles form a tetrahedron, four squares create a cube, eight triangles form an octahedron, and so on. Plato's ideas have intrigued and seduced mathematicians ever since.

In 1949, the architect and philosopher Buckminster Fuller devised his famous 'geodesic' concept. Inspired by what had once seemed no more than abstract geometry of the kind promoted by Plato, a geodesic dome is able to cover more space without internal supports than any other enclosure, by linking flat shapes together. It becomes proportionally lighter and stronger the larger it is. Fuller discovered that this abstract concept lent itself with amazing efficiency to a real-life application.

Thirty years later, Harry Kroto at the University of Sussex was studying the microwave spectra of carbon-rich stars. Until then, it was thought that pure carbon could exist only as diamond or as graphite. Kroto had discovered a third form, although he was not yet sure what kind of molecular shape it took. With help from Robert Curl and Richard Smalley at Rice University in Texas, he discovered that it was a stable molecule of exactly 60 atoms, which turned out to have a shape precisely equivalent to a geodesic dome designed by Fuller for the 1967 Montreal World Exhibition. Kroto and his colleagues named this new form of carbon 'buckminsterfullerine', or the 'buckyball'.

There are many mathematical abstractions out there. Are they inventions of the human mind, or are we indeed dimly perceiving some kind of external deep truth, which occasionally – or perhaps, always – finds its reflection in the actual structures of nature? At any rate, buckyballs are rapidly becoming the basis for new carbon-based nanotechnologies. Extremely thin carbon fibres ('nanotubes') adapted from them promise to yield structures of immense tensile strength in comparison to their size and lightness.

Plato would have considered any physical manifestation of geometry to be flawed, but buckyballs might have impressed him with their absolute, unimpeachable atomic perfection.

A computer model of a buckminsterfullerine (right) gives some impression of the molecule's regular shape. The geodesic dome designed by Buckminster Fuller for EXPO '67 in Montreal (below) has more facets, but is based on similar geometric principles.

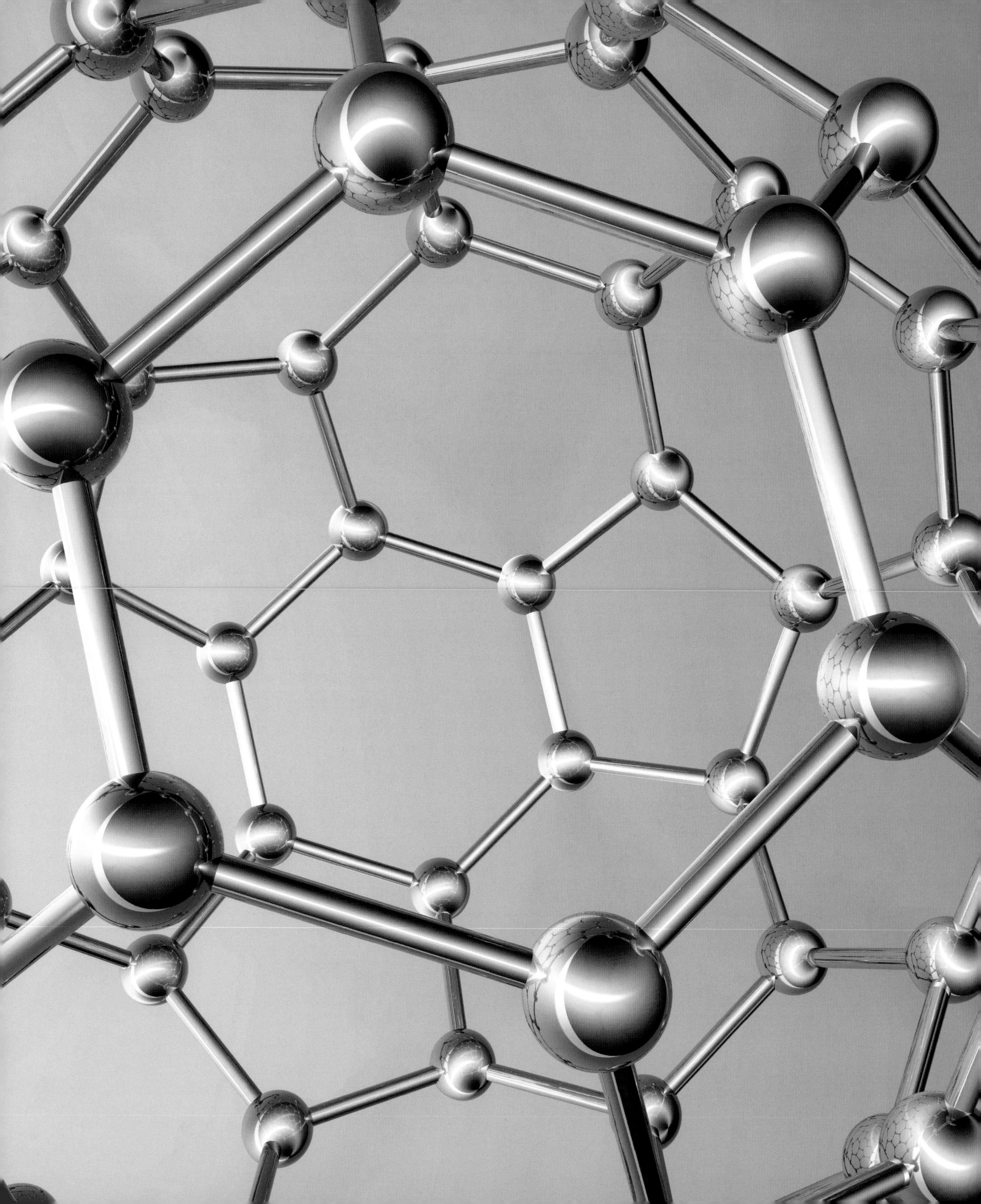

The electronic garden
simulations of natural growth

DNA is often described as the 'blueprint' for life, but this analogy is not quite right. A plan of a house or an aircraft shows all the components in relation to each other, as they should finally appear. DNA does not contain any such forward-looking template. It is more like a recipe: manipulate a given set of molecules a certain number of times in a particular way, and a complete organism should emerge.

What kind of 'language' might DNA be exploiting? It is built from just a few atoms: carbon, hydrogen, oxygen, nitrogen and phosphorus. Combinations of these form the familiar double helix backbone of the molecule, along with the compounds that carry the genetic information: thymine (T), adenine (A), cytosine (C), and guanine (G). Leaving aside rare mistakes, cytosine (C) will only pair with guanine (G), and adenine (A) only with thymine (T). When these pairs are strung together in billions-long sequences like rungs on a ladder, they are startlingly analogous to computer programs made of nothing more than '0's and '1's. In 1968, botanist Aristid Lindenmayer realised that some of the principles governing the growth of living structures could be analysed with a simple rule-based computer language. Named after the first letter of Lindenmayer's name, L-Systems is a program that simulates the shapes of trees and other plants. It can offer, perhaps, a small clue as to how DNA turns simple coding into complicated organisms.

Starting out, say, with a single vertical trunk, the topmost end might be instructed to sprout three new branches, all pointing in different directions but sharing a common root at the tip of the original trunk. Each new branch then sprouts a trio of fresh shoots, so that nine additional branches appear; and so on. After five iterations of the program, the original trunk has sprouted a canopy with 243 branches. Minor adjustments to the initial program deliver different trees: maybe a gentle curve applied to each branch when it first sprouts, or an instruction to make new branches sprout halfway along old ones. Entire tree 'species' can be stored in a few lines of computer code, just as DNA preserves an animal or plant species.

The trouble with any program based on rules is that nature as a whole often ignores the rule book. For instance, why is it that most of a real tree's branches carry on growing, but some do not? Large branches can die without falling off, or their death may be only partial, with a few shoots still sprouting from small regions where nutrients are still in supply. As trees get older, the tidy symmetry of their youth is degraded. A realistic rendition may require whole branches to be lopped off at random, as though struck by lightning – a natural event that shapes many trees, yet which is not coded into their DNA.

A stage in the construction of a virtual reality chestnut tree with autumn foliage, part of a research project conducted by the Montpelier Agricultural Research Center. The aim is to better understand how trees grow, and what factors (such as disease and climate) can affect them.

Is life a game?
cellular automata & the natural world

Mathematics can just about analyse the path of a few electrons in a supercollider; it cannot explain how a wolf works. It is an excellent tool for predicting what the universe will do, but it cannot always tell us *why* things behave as they do. It may lack something as an explanatory tool. Galileo said confidently that 'the language of nature is mathematics', but this may not be quite true. Nature may be written in simple 'rules' rather than equations, and those rules may be able to explain great complexities that conventional mathematics cannot tackle.

In 1970, Cambridge scientist John Conway invented a computer game he called 'Life'. The rules were very straightforward. Dark squares (cells) are arranged in a grid, somewhat like a chequer-board. Cells stay alive if they have an optimum number of neighbours (at least two) and die if they are left alone or overcrowded. If conditions are just right (three neighbours), they will give 'birth' to another 'live' cell nearby. The human programmer decides on an initial pattern of cells, and then the computer takes over, applying the rules again and again. Too few cells in the initial pattern, or cells spaced too widely apart, may lead to an empty grid at the end of the game, but something as simple as a 'T'-shaped initial pattern, for instance, produces amazing results. Cells breed, live and die, filling the available space in the grid with patterns of stunning orderliness and complexity. Watch the game develop, especially at high speed on a computer, and it is hard not to imagine some link with the actual workings of nature. And not an equation in sight – just the rules governing the 'on–off' digital pulses in the computer, and the even simpler rules determining how the square cells will react to each other. Even the computer is an extraneous complication, and is merely a convenient way of speeding up the experiment; 'Life' works just as well with a sheet of graph paper, a pencil and an eraser (just so long as the human pencil-pusher applies the rules consistently).

Do subatomic particles obey even simpler rules than squares on a computer screen? Presumably, a particle's repertoire of responses has to be fairly simple? Some theorists believe that the mathematics of physics and cosmology that we work with today are too complicated, and too little able to explain *why* things happen, as opposed merely to predicting *what* will happen. A rule-based theory might provide a better explanation of energy and matter, looking at the natural world, as it were, from its own point of view: the view of subatomic entities that cannot in themselves behave mathematically, yet which might be responsive to basic cues from their neighbours.

The rules of this game of 'point-mapping' are simple, yet they deliver complex results. From a few hundred points of colour distributed randomly on a grid, transmitting colour information to each other, a network of millions is formed.

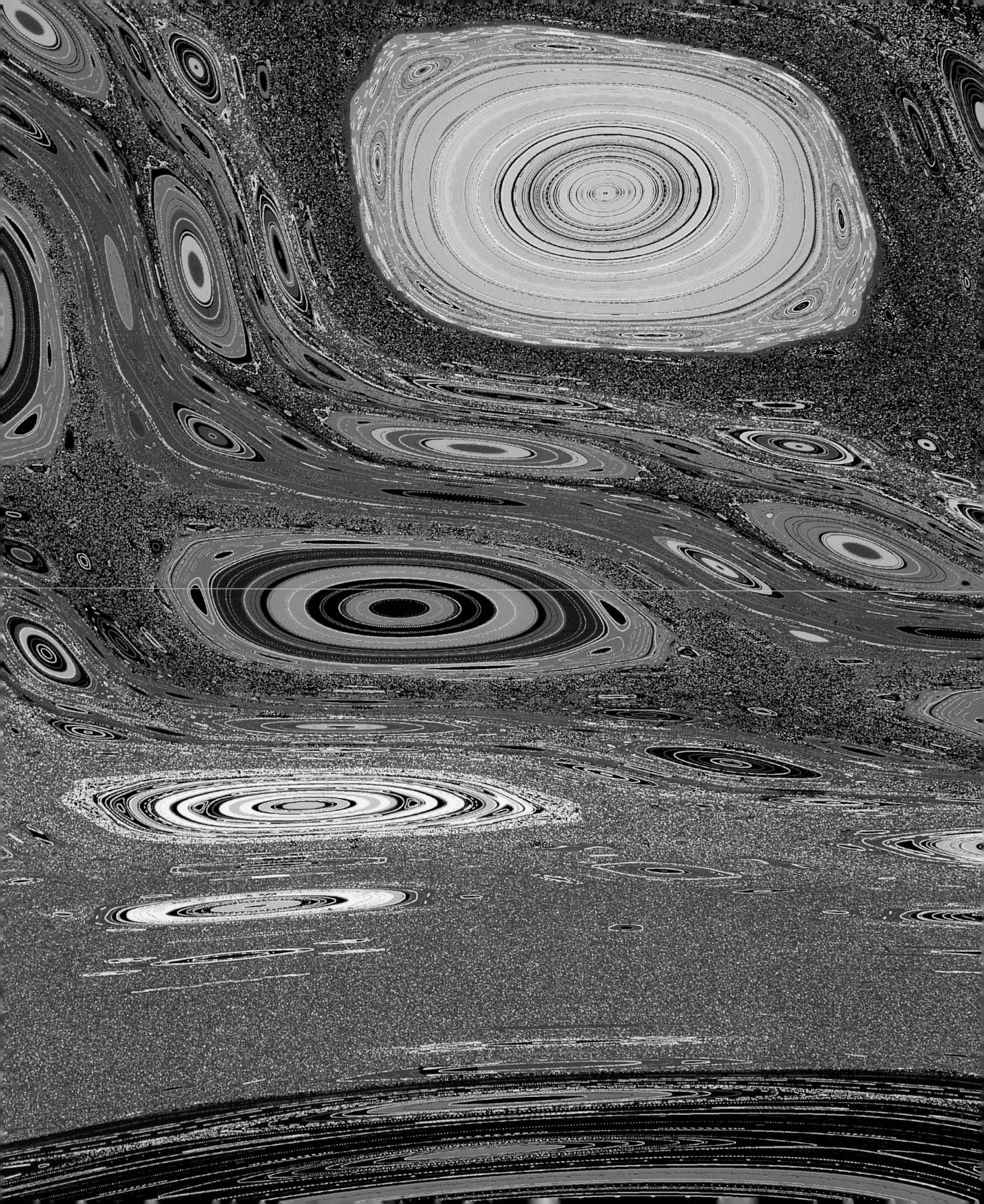

More real than life

how we judge the realism of digital effects

Digital visual effects dominate the popular movie industry. Hollywood is keen on eye-popping adventures with simplistic dialogue, so that audiences from many different nations and cultural backgrounds can follow a story. Hence the increased reliance on special effects when scripts tend to be so vapid.

As members of the audience, we are invited to marvel at the realism of, say, a miniature White House exploding, or a city falling to dust in an earthquake. Epic disasters of recent times allow us to imagine fairly well what such fictional events might look like, as we can compare the movie effects with similar realities. But what about dinosaurs that no one can ever have seen, or alien monsters that are purely imaginary? How do we judge the 'realism' of such speculative digital illusions?

Any digitised object – even the most fantastic – can be made to seem perfectly real if attention is paid to the way light appears to hit its surface, and how its shadow interplays with the background scenery. A 'virtual' lighting kit uses data collected from the physical set, where the human actors play their parts. As soon as their scene has been shot, technicians place a special ultra wide-angle electronic camera in the middle of the set, and capture the positions, colours and intensities of every light source. The camera also picks up reflected light from the walls and props of the set. This data is then 'mapped' on to the surface of an imaginary sphere. This determines how the various kinds of light should appear to strike the surface of the digital fantasy object. When that object is inserted into footage from the original set, it seems real.

Most audiences are not concerned to know the technical details, and yet if the digital lighting was wrong in even the subtlest way, they would be able to tell that something did not look quite real. On the other hand, we can be persuaded to believe in any digital illusion, so long as the joins between the real and the virtual do not show. Effects artists relish the opportunity to make wild fantasies seem real. But so far, no one has developed a computer that can visualise 'sad' or 'happy' expressions on (for instance) a mouse's face, because we have not yet created rule-based software to mimic the subtleties of emotional expression. Computer animation still requires painstaking human intervention, frame by frame.

Skilled computer animators made this forlorn mouse (he has just failed a hoop-dancing audition). The fur was created mathematically in the computer, along with the convincing effect of 'blurring' that we expect to see in an extreme close-up of a mouse.

A universe of strings
thinking beyond four dimensions

The bubble chamber images encountered earlier in this book show the tracks of myriad particles, building blocks of matter and energy at the tiniest scales. At the other extreme, space telescopes reveal the effect of gravity on vast things like planets, stars and galaxies. The challenge of physics has been to link these two ends of the scale. Gravity's detectable influence on individual particles is all but nil, yet the universe as a whole is obviously shaped by gravity. Given that the universe is made of particles, gravity must affect them somehow – and arise from them too. The search is afoot for the 'graviton', the particle that carries the force of gravity. So far, we have not found it. However, we have theorised how it might be generated.

The graviton problem is simplified if we think of particles as just the surface manifestations of something deeper. Many scientists believe that all particles of matter, and all force-carrying particles too, are generated by extremely tiny vibrating 'strings' of energy. One mode of vibration, or 'note', makes a string behave as an electron, another as a photon, and so on. Best of all, 'string theory' includes a vibration mode to deliver the graviton. At last, gravity can be described subatomically as well as at the cosmic scale. All of physics might soon be tamed so that it appears akin to harmonious music from a single stringed instrument, rather than a bewildering cacophony of noises from an argumentative orchestra. The hope is that a string-based 'theory of everything' will bind all the laws of nature into one equation. It will explain the entire universe as surely as it can explain a single atom.

One snag: strings are much smaller than particles, so we have no obvious prospect of detecting them (or, consequently, proving their existence) with even the most advanced instruments that study particles. Another difficulty: while the mathematics of string theory seem robust, they require outlandish assumptions. We experience the world in four dimensions: one in time, and three in space. Strings supposedly exist in nine dimensions of space. Six hidden dimensions are wrapped up in invisibly tiny bundles, associated with the strings, called Calabi-Yau Manifolds. However, this may be a convenient method of sweeping the excess dimensions under the carpet, where no one has to worry about them in practical terms. They are there, the theory says, but so tightly squeezed together we have no way of observing them.

A computer visualisation of the almost unimaginable: strings of energy vibrating at infinitesimally tiny scales, with wave harmonics producing the particles and measurable energies that we are able to observe in our 'real' world.

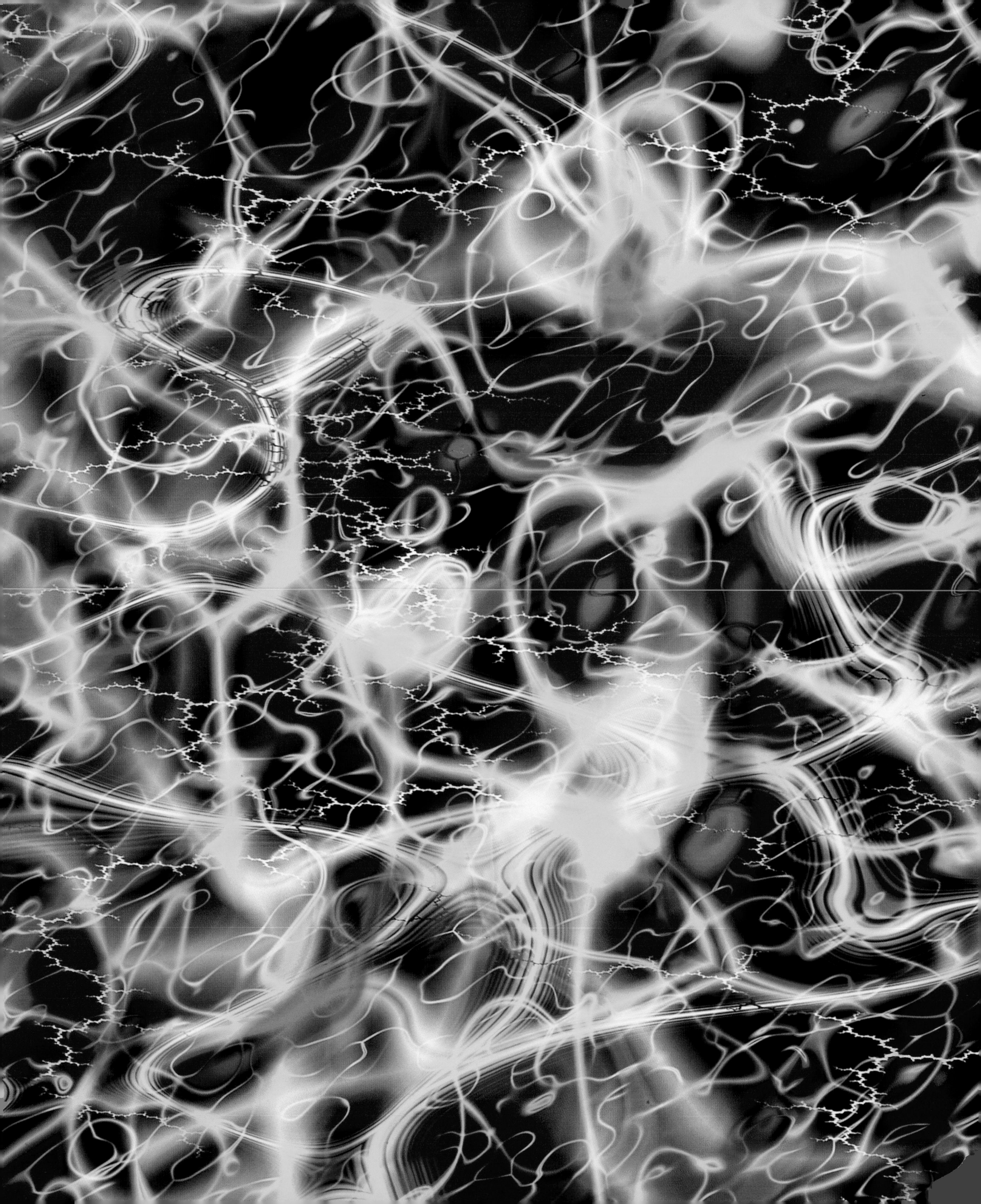

The multiverse

is there more to existence than this one universe?

Twenty years ago, the idea of multiple universes seemed a fantasy. Now it threatens to become scientifically necessary. The existence of other universes is logically implied by some of the theories that best explain our own. The six extra dimensions of string theory, for instance, have to be wished away by bundling them up very tightly, but it is conceivable that other universes, built by the same strings, conceal our spatial dimensions in a similar way. Quantum theory shows that subatomic particles can appear as discrete objects when we observe them, but the rest of the time (when we are not looking) they are a ghostly cloud of probabilities and potentialities. All the theoretically allowable outcomes of their behaviour exist as a matrix of possibilities called 'superposed' states. When we at last observe the particles by making a scientific measurement, their superpositions collapse into just one specific state, and we record 'definite' particle events and qualities. We build up a history, through time, of 'what has happened' in our world. Observations shape our reality, but all the different particle events (superpositions) that we *might* have observed are no less real – it is just that they happen in parallel universes that split away from ours the instant we make those observations. With every passing moment, alternate realities accumulate in the quantum 'multiverse'. Somewhere, another you exists that has not chosen to read this book.

In April 2003, the *New York Times* ran an article by the Australian physicist Paul Davies warning readers not to take multiple universes seriously. He argued that it is not good science to speculate about things that are intrinsically unobservable, because there is no way to prove them. The next month, *Scientific American* published an article by physicist Max Tegmark asserting that parallel universes almost certainly must exist.

This universe seems improbably fine-tuned for life. The Big Bang should have been able to deliver an infinite variety of less successful universes, but it came up with this one, first time out of the hat. Many cosmologists therefore think that some overarching intelligence fine-tuned the forces of nature. However, some thinkers want to turn the problem on its head. They insist that other universes must exist precisely to make superstitious mysteries go away. It would make more sense if big bangs were generated in profusion, just like stars and galaxies. That would leave room for some bangs to succeed, and others to fail. Nature spawns countless universes unthinkingly, and leaves them to succeed or fail on their own terms. They exist as a vast population, but their only linkages are through exotic wormholes in space-time, or black holes and other such extreme phenomena.

The late Martin Gardner, renowned for his mathematical inventiveness, said, 'Surely the conjecture that there is just one universe and its Creator is infinitely simpler and easier to believe than that there are countless billions upon billions of worlds?'

For the time being at least, Gardner's assertion remains a matter of opinion.

A speculative image showing multiple universes splitting off from each other in the wake of the Big Bang. They become a population of universes, a 'multiverse' scattered across multiple dimensions, unable to communicate with each other except, possibly, through wormholes.

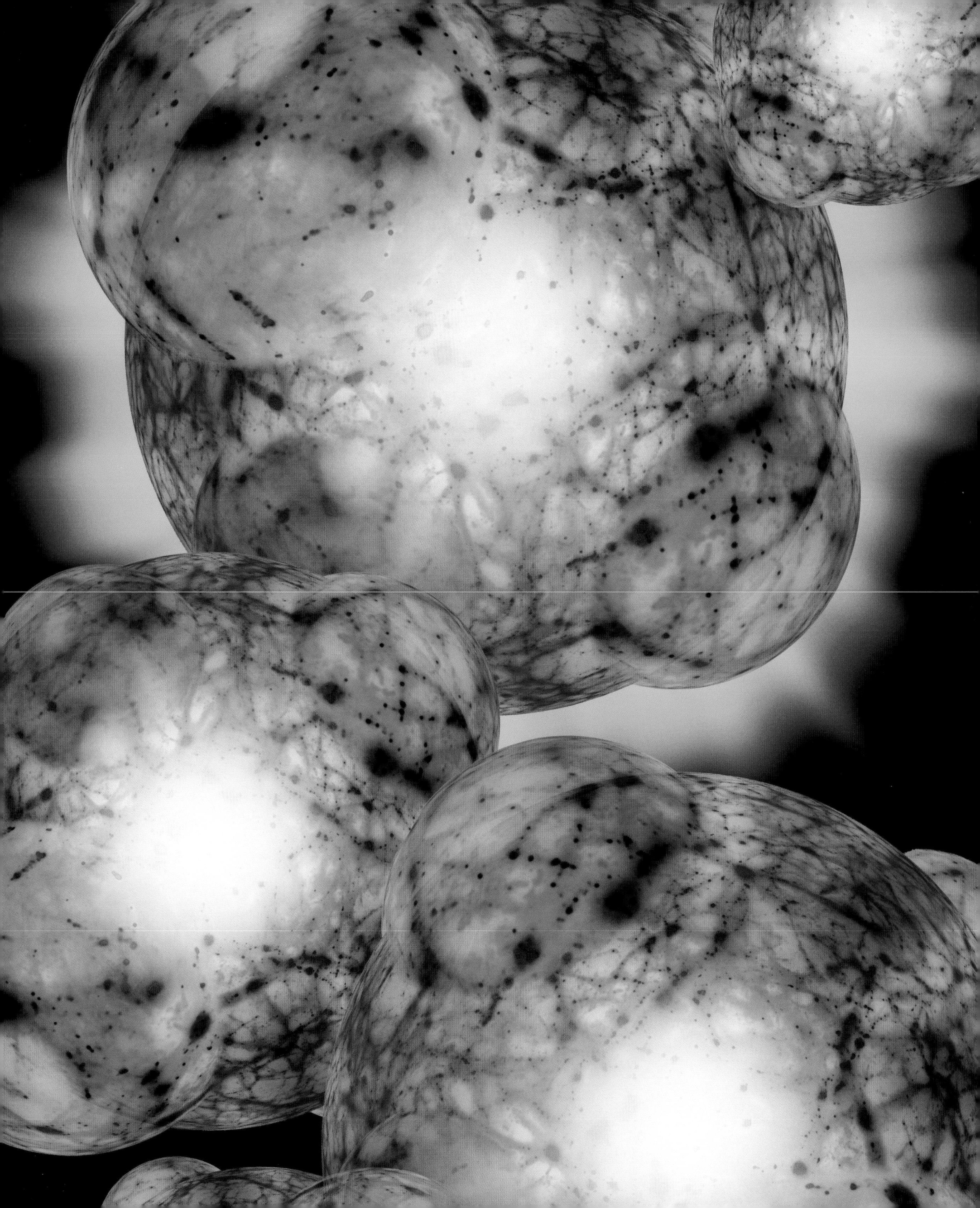

Is reality beautiful?

the temptations of symmetry & neatness

The human brain is attuned to natural orderliness for good evolutionary reasons. A round apple is likely to be better than a bumpy one, because bumps signify disease or insect blight. A fallen plum, bruised and covered in dirt, is less enticing than a smooth clean one still hanging from its branch. A fish with scabby scales and one eye does not make an appealing catch. Well-balanced features even influence our selection of partners, because symmetrical bodies tend to signify healthy genes. These are simplistic examples, but evolution has driven us to take pleasure in beauty, balance, harmony and other such aesthetically pleasing qualities because they usually imply something beneficial for us as animals.

So what happens if our instinctive desire for beauty becomes intellectually abstract rather than purely survivalist? Then we encounter scientists saying that an elegant equation 'feels' just right and 'fits' neatly with their observations of the real world.

Scientists in general have a love of beautiful equations, and tend perhaps too easily to think that they must imply truth. In May 1963, Nobel laureate Paul Dirac wrote in *Scientific American*, 'It is more important to have beauty in one's equations than to have them fit experiment... If there is not complete agreement between the results of one's work and experiment, one should not allow oneself to be too discouraged, because the discrepancy may well be due to minor features that are not properly taken into account and that will get cleared up with further development of the theory.' The mathematician Godfrey Hardy said in 1941, 'Beauty is the first test. There is no permanent place in this world for ugly mathematics.' Albert Einstein was never lost for words on the subject of his fascination for pretty equations: 'It is possible to know when you are right way ahead of checking all the consequences. You can recognise truth by its beauty and simplicity.'

Great intellects are inclined to reject mathematical apples if they look rotten, but in nature rotten apples are just as real as healthy ones. Today, some scientists believe we should pay a little less attention to things that can be interpreted with neat equations. Instead, we should look more closely at the messy results, the ugly, inconsistent data that will not necessarily please our instincts. Beauty may be less reliable a guide to truth than we care to realise.

Most of the pictures in this book are of limited value to scientists – they are approximations and decorations, often not as important as numerical data. The fundamental tool of science is mathematics. Here we see what cutting-edge physics 'really' looks like.

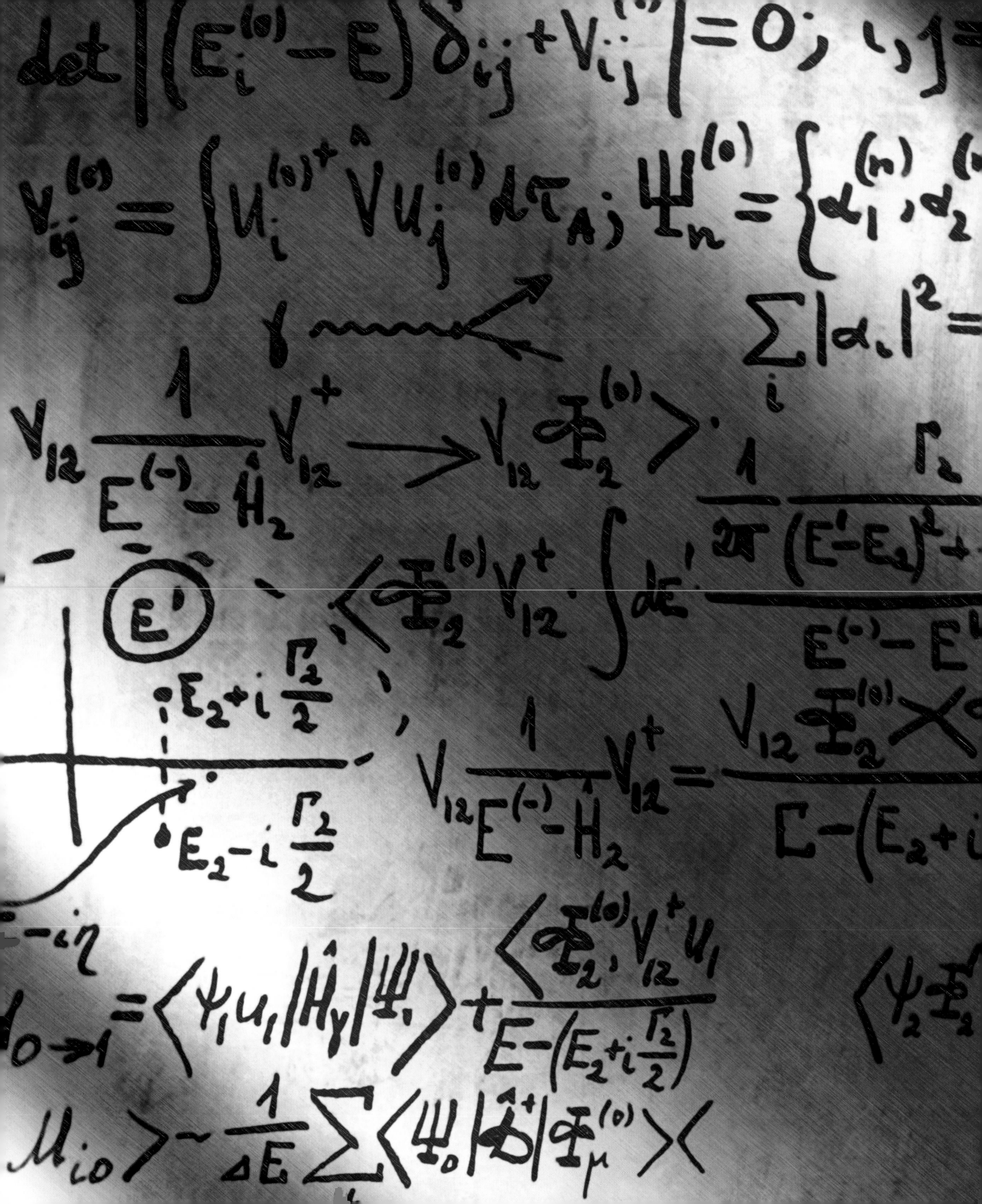

What does it mean to see?

the 'qualia' of human vision

Colours do not exist. Nature generates electromagnetic waves, and our retinas respond electrochemically in certain consistent ways, but colours happen purely inside the perceptual workings of the brain. They have no reality outside our skulls. Colour blindness charts prove that we can all distinguish between colours, because it is not the colours themselves so much as a pattern of numbers or letters hidden in a jumble of multicoloured dots that settles the issue. We know that people with normal eyesight do not mistake red for green or blue for yellow. That being the case, as long as we can agree that red 'means' the same thing to different people, then it may not be terribly important whether or not we actually *see* red in the same way as others do. We can share the cultural and historical associations of 'blue' and 'red' but cannot be sure we are experiencing the same colour sensations in our minds. Our awareness of colour is private; perhaps terrifyingly so, if brooded on. It is also arbitrary, because the brain perceives different versions of the same colour to suit its expectations. A pink card held up against a red background will look redder when held against a blue background. In a room with pink lighting and pink walls, we might have difficulty telling a white card from a pink one of the same hue as the walls. We perceive colours in the context of other colours around them. The wavelengths of the spectrum are absolute, and can be measured precisely by dispassionate instruments – but our personal experiences of colour are subjective.

We read in a book that a healthy apple should be green with a blushing red patch. This is knowledge that helps us pick good apples when we go to the store, but it is not the same as becoming aware of the redness and greenness once we have the apple in our hand. When the colours come alive in our minds, or when we recall, later on, what those colours looked like, we are experiencing a phenomenon known as 'qualia', the most vivid form of consciousness. A supercomputer could tell ripe apples from unripe ones by measuring the wavelengths of light bouncing off them, but could it *experience* the greenness of things as we can? These are deep questions about how we see the world.

We have learned to access all regions of the spectrum and turn them into qualia of sorts. In practice, this comes down to translating non-optical wavelengths into those that we can see. After a while, though, looking at these miracles of translation, one might begin to get a little fed up with the limitations of blue, green, yellow and red in the rendered results, when there are so many other kinds of light out there, if only we could see them properly.

This is a standard colour-blindness chart. You should be able to discern a two-digit number inside the disc, formed by dots markedly different in colour from the background.

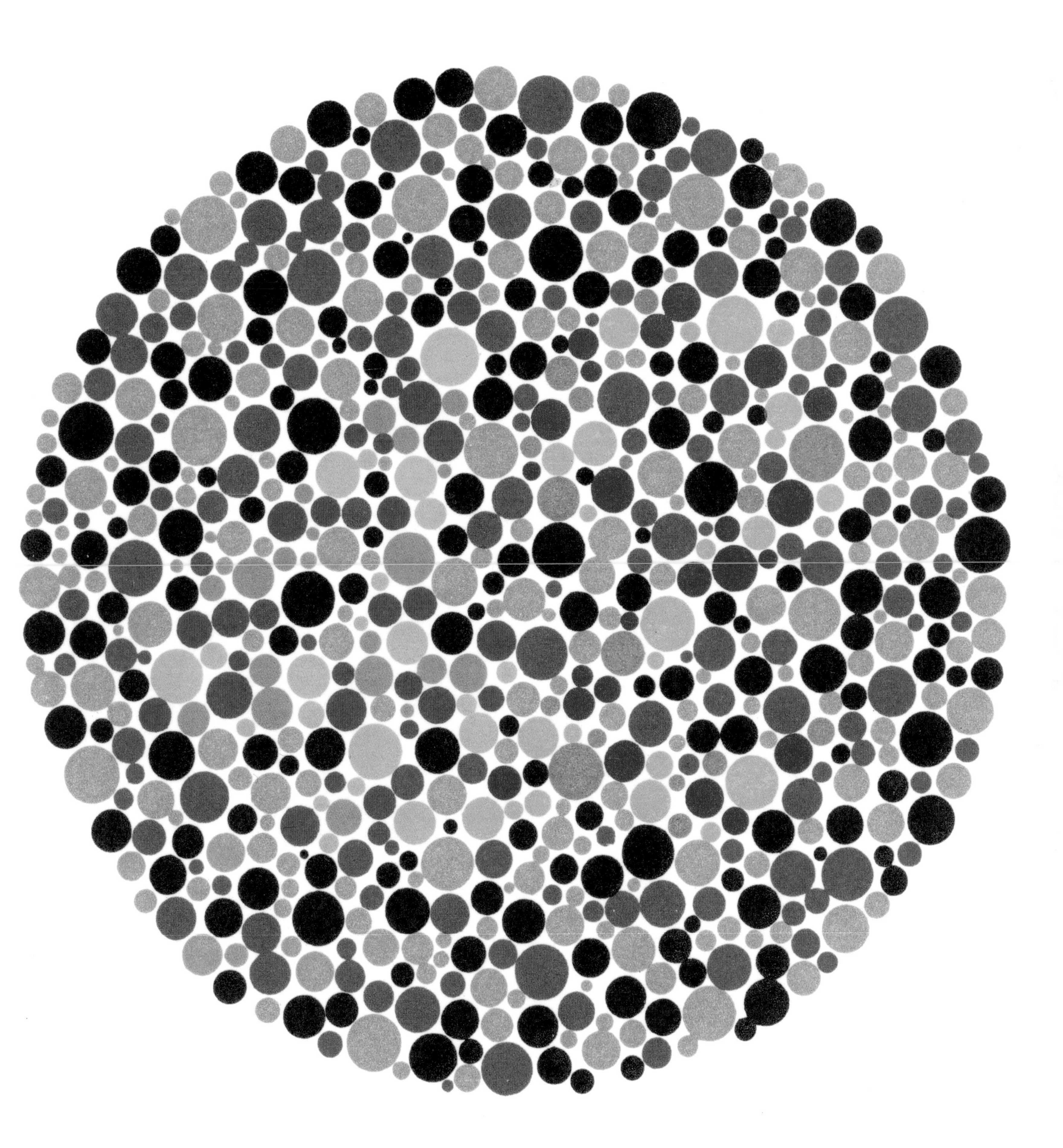

The universal computer
is reality an illusion?

Physics seems to be showing us that everything we know of, from cats, dogs, trees and people to stars, planets and galaxies, is all made from exactly the same fundamental building blocks. Science is closing in on a set of equations which can account for everything using a few simple rules. We might then be left with a more subtle question: how do we account for the differences we observe between things, when they are all made of the same kind of stuff?

One possible answer is that the cosmos behaves like a computer, 'outputting' the version of reality that we observe, rather than actually 'being' that reality. It uses a code of subatomic components and forces, and combines them according to the rules of its software (the 'laws of physics') to deliver many different outputs, just as an electronic computer manipulates simple '0's and '1's to create, for instance, complex images on a screen. Scientists are now debating the potential importance of information as the true mechanism at the heart of existence. Any subatomic sample of the cosmos is all but indistinguishable from another. Pull back to the atomic and macroscopic scales, though, and distinct chemical structures begin to emerge. Pull back further, and the cosmos seems suddenly rich in its variety of objects and events. Yet they may all be nothing more than shimmering assemblages of subatomic information – and as malleable and essentially impermanent as any computer's temporarily stored patterns of digital data.

One key point about a computational system (as we understand it) is that the information has to be manipulated by some kind of surrounding physical architecture: a modern computer chip, for instance, with its countless 'on-off' logic gates and memory modules for storing the results of its calculations. There is some confusion among information theorists about whether or not information has any real meaning outside the architecture that manipulates it. If the cosmos is similarly information-based, the question arises: what is the hidden architecture that stores all the information, manipulates it and outputs the results? From our point of view, observing a particle in an accelerator, or through a scanning tunnelling electron microscope, is roughly the equivalent of registering the tiny fluctuation of shifting energies that occurs, say, as a computer memory chip transistor flicks from '0' electric charge to '1'. We can observe the local effects of little pieces of information in flux, but we will need a much wider view if we are to discern the broader processing architecture of the cosmos, and the greater matrix of information that creates our reality.

An architectural model of an office building is displayed on the inside of a spherical auditorium. In the near future, computer simulations like this will be indistinguishable from reality. Our perceptions of the universe as a whole may be a similar illusion.

Predicting the apocalypse

visualising our possible end

An Assyrian clay tablet dating from 3,000 years before the birth of Christ bears the words: 'Our earth is degenerate in these latter days. There are signs that the world is speedily coming to an end.' Pope Innocent III expected Armageddon to take place in the year 1284, exactly 666 years after the rise of Islam. The Czech doomsday specialist Martinek Hausha warned that the world would be destroyed in February 1420 (14 February at the latest, he specified). Throughout history, priests, soothsayers, charlatans and opportunists have predicted our end with gleeful enthusiasm.

Modern scientists have their doomsday scenarios too, based on somewhat better theories than those of ancient astrologers and mystics, yet still flavoured with guesswork and speculation. The millennium celebrations for the year 2000 were attended by a fear of global meltdown caused by overloaded computer chips. Geologists have warned that Vesuvius, the volcano that destroyed Pompeii in 79 AD, is about to strike again, threatening the lives of millions of people in towns around the Bay of Naples. Global warming experts predict widespread flooding of coastal regions some time in the next generation or two. Other climatologists believe that the increased cloud cover caused by global warming will work a more subtle disaster. So much sea water will evaporate in the heat, the entire planet will be girthed in cloud all year round; temperatures will then fall, generating a new ice age, killing off most complex life on the planet. The Yellowstone National Park in America sits on top of a crustal plate which is being pushed upwards by a magma surge of unimaginably huge proportions. If there is an eruption, the smoke and dust cloud will be so huge it could blot out the sun for a generation, killing off . . .

To cap it all, asteroid-watching space agencies have calculated that a major city-destroying meteorite should slam into us once every 100,000 years, and an 'extinction-level' impact should occur once every 50 million years. We may be overdue for both city-smashing and extinction-level impacts.

In 1908, a 60-metre wide meteorite blew up in the air above the remote Siberian forest region of Tunguska, with the force of 15 million tons of TNT. Hundreds of square kilometres of trees were flattened like matchsticks. There were only a few dozen eyewitnesses, loggers and peasants whose reports were not taken seriously by the authorities in Moscow. A full scientific survey of the area was not conducted until after the First World War. However, the vast flattening of trees, in a strange butterfly-wing pattern, was eventually mapped. Today, our estimates of the size and power of the Tunguska object are based on sophisticated computer simulations. It turns out that stray asteroids just a few tens of metres across, streaking obliquely across the planet at low altitude, could wreak more damage than an atomic bomb. The solar system is criss-crossed by countless thousands of chunks of wayward rubble, many of them several kilometres in diameter.

End-of-the-world scenarios continue to exert a strange and terrible fascination. Now we have the tools to 'see' the coming disaster of our choice.

A simulation of a massive 'extinction-level' asteroid strike on the Earth. We can reasonably expect a catastrophe of some kind along these lines: perhaps the destruction of a city, or a tidal wave caused by an impact in the ocean.

©D. van Ravenswaay '98

Our quest to be seen

messages sent into space

All the probes we send into the cosmos are designed to explore an alien realm, but only one was intended specifically to communicate with aliens. Launched on 2 March 1972, Pioneer 10 was the first spacecraft to obtain close-up images of Jupiter. Plutonium batteries kept the tiny craft operating far longer than anyone had anticipated. Its last, almost imperceptibly faint signal was returned in January 2003, more than 30 years after its launch. By then, the entire solar system was 13 thousand million kilometres in its past.

Mission planners always knew that Pioneer 10's trajectory would eventually fling it away from the Sun and out towards the stars – specifically, towards Aldebaran in the constellation of Taurus, 68 light years away. Pioneer will not reach Aldebaran for another 2 million years, but its ghostly shell might survive the trip: a fossil remnant of an earthly civilisation perhaps long vanished by then. There is a slight chance that its artificial construction will be noted by an alien intelligence in the depths of space.

In the early 1970s, a plaque was designed to attach to the probe's side, made from corrosion-proof gold and aluminum. It shows where our species lives, and its biological form. At top left is a schematic of the hyperfine transition of neutral atomic hydrogen, supposedly a universal 'yardstick' providing a basic unit of both time and physical length throughout the universe. As a further size check, the binary equivalent of the number '8' is shown between tote marks indicating the height of the two human figures, to be compared with the scale of the spacecraft itself, which is also shown in line silhouette on the plaque. The radial pattern to the left represents the position of the Sun relative to 14 pulsars and to the centre of the Galaxy. The binary digits on the other lines denote time. There is also a diagram showing Pioneer 10's track, accelerating past the largest planet in our solar system with its antenna pointing back to its origin, the third planet.

Would you understand the plaque, as drawn by people of your own species? Perhaps not at first, but intelligent aliens might dedicate their brightest minds to the task of unravelling its mysteries, so maybe they will have better luck.

For now, the plaque is better understood (by us, at any rate) as a message to ourselves rather than to aliens. It signifies that we at least tried to head towards the stars even while clashing ideologies and social problems afflicted us on Earth. It is a technological cave painting, created as much because we felt like it as for any other reason; a general portrait of ourselves as we would like to be seen. At least, it *should* be, but arguments have raged since 1972 about the assumptions made in the scientific symbols, the arbitrary ethnicity of the human figures and the woman's smaller size compared with the man's.

This book has been about our new methods for seeing the world. A plaque on the side of a tiny space probe signifies our desire to be seen. We look for the traces of other intelligences in the universe, and hope that they, in turn, might be looking for us.

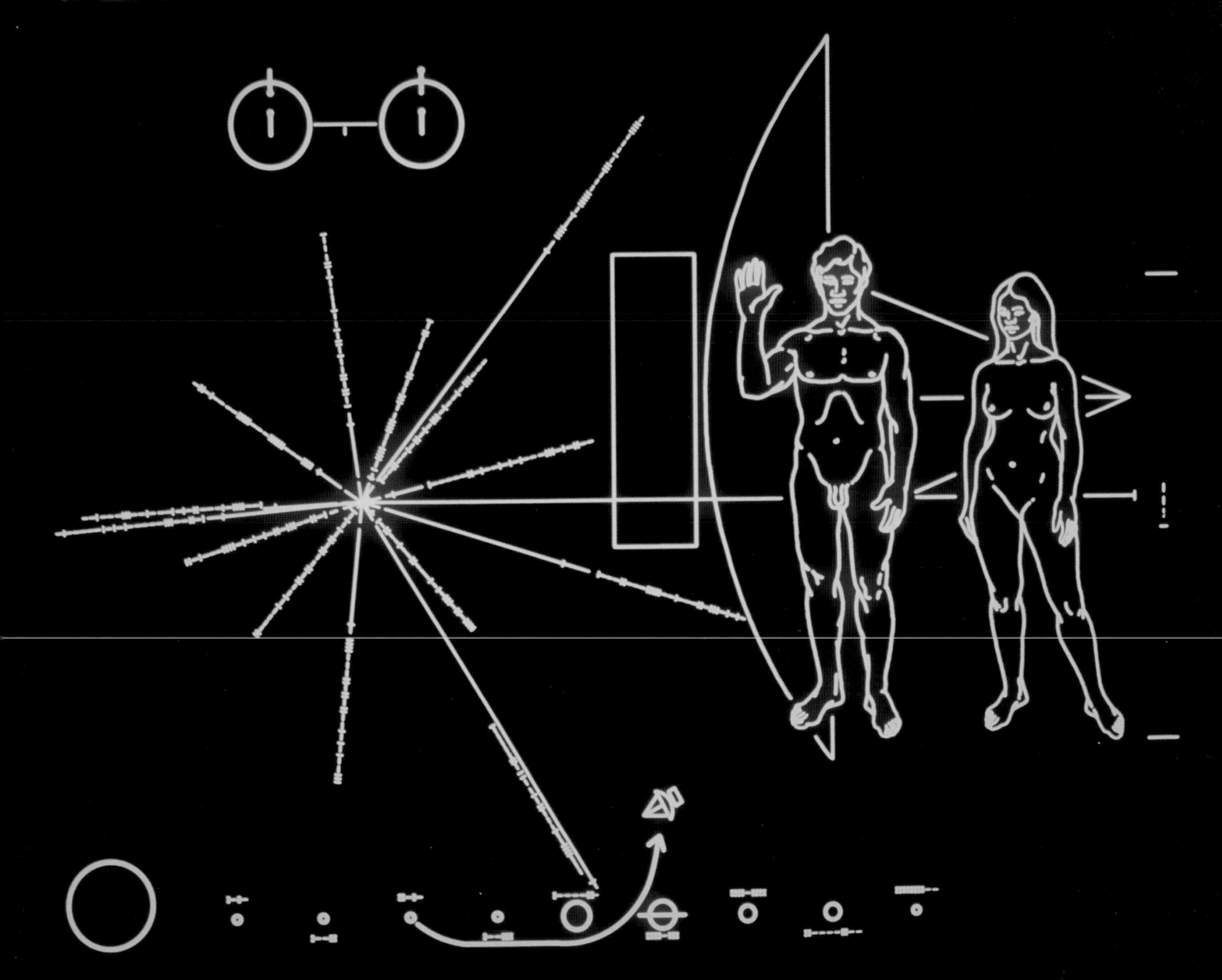
NASA

Index

Acknowledgements

The author would like to thank Dennis Gilliam and Simon Atkinson for their invaluable descriptions of space imaging systems. Monica Grady at the Natural History Museum in London is always very helpful when it comes to meteorite impacts and other geological issues. Rita Carter explained medical scanning techniques, and made some valuable contacts for me, for which, many thanks indeed. Don Savage and Roger Launius at NASA were very generous with their time. My main collaborators have been the team at Weidenfeld. In particular I would like to thank Jennie Condell, Michael Dover and Nic Cheetham for their sympathetic editorship, and Annabel Merullo for her tremendous job of picture research. Finally, my consultant editor Dr Jim Al-Khalili was always ready to put complex scientific ideas into plain English for me. Any errors in this book are the author's, not Jim's.

Picture credits

The publishers would like to thank the following individuals and institutions for permission to reproduce the pictures on the pages listed below. Every effort has been made to trace the copyright holders. Weidenfeld & Nicolson apologise for any unintentional omissions and, if informed of such cases, shall make corrections in any future edition.

2–3 NASA/JPL/University of Arizona/Los Alamos National Laboratories
4–5 Susumu Nishinaga/Science Photo Library
6–7 Marc Imhoff NASA/GSFC, Christopher Evidge NOAA/NDGC, Craig Mayhew and Robert Simmon NASA/GSFC
16 left copyright University of Liverpool; right NASA/ Goddard Space Flight Center/SPL
17 left NASA/GSFC/METI/ ERSDAC/JAROS; right NRAO/AVI/NSF/SPL
19 Patrice Loiez/CERN/SPL
21 Cordelia Molloy/SPL
23 Alfred Pasieka/SPL
25 Astrid & Hanns-Frieder Michler/SPL
27 OMIKRON/SPL
29 Lawrence Berkeley Laboratory/SPL
31 L. Medard Eurelios/SPL
33 David Parker/SPL
35 Patrice Loiez/CERN/SPL
37 Fermilab/SPL
39 CERN/SPL
41 Manfred Kage/Oxford Scientific Films
43 Alfred Pasieka/SPL
45 Dr. Mitsuo Ohtsuki/SPL
47 Erich Schrempp/SPL
48 Courtesy: IBM Research, Almaden Research Center. Unauthorised use not permitted
49 Courtesy: IBM Research, Almaden Research Center. Unauthorised use not permitted
51 M. W. Davidson/SPL
52 Courtesy: IBM Research, Almaden Research Center. Unauthorised use not permitted
53 Sandia National Laboratories/SPL
57 A. B. Dowsett/SPL
59 Dr Linda Stannard, UCT/SPL
61 EYE OF SCIENCE/SPL
63 Nancy Kedersha/Immunogen/SPL
65 David Scharf/SPL
66 Eye of Science/SPL
67 David Scharf/SPL
68 Eye of Science/SPL
69 Eye of Science/SPL
70 Andrew Syred/SPL
71 Andrew Syred/SPL
73 National Cancer Institute/SPL
74 VVG/SPL
75 VVG/SPL
77 NASA/SPL
79 Alfred Pasieka/SPL
80 SPL
81 CMRI/SPL
82 Goya, *Doña Isabel de Porcel* © The National Gallery, London
83 Goya, *Doña Isabel de Porcel* © The National Gallery, London
85 Du Cane Medical Imaging Ltd./SPL
87 Smithonian Institute/SPL
89 copyright University of Liverpool
91 SPL
93 James King-Holmes/SPL
95 Simon Fraser/SPL
97 Zephyr/SPL
98 MITAI LAB/ Surgical Planning Laboratory/Brigham's Women's Hospital/SPL
99 AGLIOLO/SPL
101 Biophotos Associates/SPL
103 Tim Beddow/SPL
105 Institute of Mathematics & Computer Science in Medecine, University of Hamburg
107 Philippe Plailly/EURELIOS/SPL
109 Mark Maio/King-Holmes/SPL
111 VVG/SPL
112 Dr. Jermey Burgess/SPL
113 Leonard Lessin/SPL
115 Alfred Pasieka/SPL
116 Michel Viard, Peter Arnold Inc.
117 Garion Hutchings/SPL
119 Dr. Arthur Tucker/SPL
120 Reynaldo Gomez, Edward Ma, Thomas Wey, NASA Johnson Space Center
121 L. Weinstein NASA/SPL
123 D. Roberts/SPL
125 Copyright American Science & Engineering, Inc.
126 Maximilian Stock Ltd./SPL
127 Walter Bibikow/Image Bank/Getty Images
129 Adam Gault/Digital Vision
131 Detlev Van Ravenswaay/SPL
133 Victor de Schwanberg/SPL
135 NASA/SPL
137 Courtesy NASA/JPL-Caltech
139 NASA/SPL
141 Aviation Images
143 W.T. Sullivan III & Hansen, Plantarium/SPL
145 NASA/SPL
147 NASA/GSFC/LARC/JPL, MISR Team
148 NASA/JPL
149 NASA/Goddard Space Flight Center/SPL
150 US Naval Postgraduate School and Mississippi State University's GRI, in conjunction with ONR/HPCMP/ NSF/DOE/NOAA
151 Dr Ken MacDonald/SPL
152 Perenco UK Ltd./PGL/Landmark Graphics
153 Geological Survey of Canada/SPL
155 SPL
157 NASA/GSFC/METI/ERSDAC/ JAROS & US/Japan Aster Science Team
159 Mark Simon/Californian Institue of Technology, 1997/SPL
161 A. Burrows/Arizona University/SPL
163 R. B. Husar/NASA/SPL
165 NASA/SPL
167 Astrogeology Team, US Geological Survey, Flagstaff, Arizona
169 Astrogeology Team, US Geological Survey, Flagstaff, Arizona
171 Detlev van Ravemswaay/SPL
173 NASA/SPL
175 NASA/SPL
176 NASA/TRACE
177 RASA/Institute for Solar Physics
179 Mathew Bate/SPL
181 Konstantinos Kifonidis/SPL
182 NASA/STScI
183 NCSA/University of Illinois/SPL
185 NASA/Goddard Space Flight Center
187 Enrico Costa, Angelo Antonelli and the BeppoSAX GRB team
188 B. Cooper & D. Parker/SPL
189 NRAO/AUI/NSF/SPL
191 NASA/SPL
193 Max Planck Institute of Aerophysics/ SPL
195 Mehav Kulyk/SPL
196 James Robinson/OSF
197 Martin Bondi/SPL
199 Gregory Sams/SPL
200 NASA/SPL
201 David Taylor/SPL
202 Jimmy Fox/SPL
203 Alfred Pasieka/SPL
205 Hubert Raguet/Eurelios/SPL
207 Eric Heller/SPL
209 Glassworks
211 Mehav Kulyk/SPL
213 Detlev Van Ravenswaay/SPL
215 Novosti Photo Library/SPL
219 Colin Cuthbert/SPL
221 D. Van Ravensway/SPL
223 NASA/SPL

First published in the United Kingdom in 2004
by Weidenfeld & Nicolson,
a division of the Orion Publishing Group

Printed and bound in Italy by
Printers SRL and LEGO

A CIP catalogue record for this book is available from the British Library.

Design director David Rowley
Designed by Ken Wilson
Edited by Jennifer Condell
Picture research by Annabel Merullo

ISBN 0 297 84342 7

Weidenfeld & Nicolson
The Orion Publishing Group
Wellington House
125 Strand
London WC2R 0BB